Ergebnisse der Anatomie und Entwicklungsgeschichte
Advances in Anatomy, Embryology and Cell Biology
Revues d'anatomie et de morphologie expérimentale
Springer-Verlag · Berlin · Heidelberg · New York

This journal publishes reviews and critical articles covering the entire field of normal anatomy (cytology, histology, cyto- and histochemistry, electron microscopy, macroscopy, experimental morphology and embryology and comparative anatomy). Papers dealing with anthropology and clinical morphology will also be accepted with the aim of encouraging co-operation between anatomy and related disciplines.

Papers, which may be in English, French or German, are normally commissioned, but original papers and communications may be submitted and will be considered so long as they deal with a subject comprehensively and meet the requirements of the Ergebnisse.

For speed of publication and breadth of distribution, this journal appears in single issues which can be purchased separately; 6 issues constitute one volume.

It is a fundamental condition that manuscipts submitted should not have been published elsewhere, in this or any other country, and the author must undertake not to publish elsewhere at a later date.

25 copies of each paper are supplied free of charge.

Les résultats publient des sommaires et des articles critiques concernant l'ensemble du domaine de l'anatomie normale (cytologie, histologie, cyto et histochimie, microscopie électronique, macroscopie, morphologie expérimentale, embryologie et anatomie comparée. Seront publiés en outre les articles traitant de l'anthropologie et de la morphologie clinique, en vue d'encourager la collaboration entre l'anatomie et les disciplines voisines.

Seront publiés en priorité les articles expressément demandés nous tiendrons toutefois compte des articles qui nous seront envoyés dans la mesure où ils traitent d'un sujet dans son ensemble et correspondent aux standards des «Résultats». Les publications seront faites en langues anglaise, allemande et française.

Dans l'intérêt d'une publication rapide et d'une large diffusion les travaux publiés paraitront dans des cahiers individuels, diffusés séparément: 6 cahiers forment un volume.

En principe, seuls les manuscrits qui n'ont encore été publiés ni dans le pays d'origine ni à l'étranger peuvent nous être soumis. L'auteur d'engage en outre à ne pas les publier ailleurs ultérieurement.

Les auteurs recevront 25 exemplaires gratuits de leur publication.

Die Ergebnisse dienen der Veröffentlichung zusammenfassender und kritischer Artikel aus dem Gesamtgebiet der normalen Anatomie (Cytologie, Histologie, Cyto- und Histochemie, Elektronenmikroskopie, Makroskopie, experimentelle Morphologie und Embryologie und vergleichende Anatomie). Aufgenommen werden ferner Arbeiten anthropologischen und morphologisch-klinischen Inhaltes, mit dem Ziel die Zusammenarbeit zwischen Anatomie und Nachbardisziplinen zu fördern.

Zur Veröffentlichung gelangen in erster Linie angeforderte Manuskripte, jedoch werden auch eingesandte Arbeiten und Originalmitteilungen berücksichtigt, sofern sie ein Gebiet umfassend abhandeln und den Anforderungen der „Ergebnisse" genügen. Die Veröffentlichungen erfolgen in englischer, deutscher oder französischer Sprache.

Die Arbeiten erscheinen im Interesse einer raschen Veröffentlichung und einer weiten Verbreitung als einzeln berechnete Hefte; je 6 Hefte bilden einen Band.

Grundsätzlich dürfen nur Manuskripte eingesandt werden, die vorher weder im Inland noch im Ausland veröffentlicht worden sind. Der Autor verpflichtet sich, sie auch nachträglich nicht an anderen Stellen zu publizieren.

Die Mitarbeiter erhalten von ihren Arbeiten zusammen 25 Freiexemplare.

Manuscripts should be addressed to/Envoyer les manuscrits à/Manuskripte sind zu senden an:

Prof. Dr. A. BRODAL, Universitetet i Oslo, Anatomisk Institutt, Karl Johans Gate 47 (Domus Media), Oslo 1/Norwegen.

Prof. W. HILD, Department of Anatomy, The University of Texas Medical Branch, Galveston, Texas 77550 (USA).

Prof. Dr. R. ORTMANN, Anatomisches Institut der Universität, 5 Köln-Lindenthal, Lindenburg.

Prof. Dr. T. H. SCHIEBLER, Anatomisches Institut der Universität, Koellikerstraße 6, 87 Würzburg.

Prof. Dr. G. TÖNDURY, Direktion der Anatomie, Gloriastraße 19, CH-8006 Zürich.

Prof. Dr. E. WOLFF, Collège de France, Laboratoire d'Embryologie Expérimentale, 49 bis Avenue de la belle Gabrielle, Nogent-sur-Marne 94/France.

Ergebnisse der Anatomie und Entwicklungsgeschichte
Advances in Anatomy, Embryology and Cell Biology
Revues d'anatomie et de morphologie expérimentale

40 · 6

K. France Baker-Cohen

Comparative Enzyme Histochemical Observations on Submammalian Brains

Part I. Striatal Structures in Reptiles and Birds

With 41 Figures

Part II. Basal Structures of the Brainstem in Reptiles and Bird

With 51 Figures

Springer-Verlag Berlin Heidelberg GmbH 1968

K. France Baker-Cohen, Ph. D., Assistant Professor of Anatomy
Department of Anatomy
Albert Einstein College of Medicine, Bronx, New York 10461, USA

Supported by Training Grant NIH-5 T1-GM-102 from the
National Institutes of Health, Public Health Service, Bethesda, Maryland

ISBN 978-3-540-04090-3   ISBN 978-3-642-85941-0 (eBook)
DOI 10.1007/978-3-642-85941-0

Library of Congress Catalog Card Number 64—20582 Titel-Nr. 6954. Printed in Germany

Contents Part I

Introduction 7

Materials and Methods. 8

Observations and Comparisons 10
    Palaeostriatum 10
    Dorsal Striatum 15

Discussion 31

Summary 38

Bibliography 39

Contents Part II

Introduction 42

Materials and Methods. 42

Observations 43

A. Ventral Thalamus and Subthalamus 43
    1. Nucleus of the Dorsal Supraoptic Decussation 43
    2. Entopeduncular Nucleus 57

B. Midbrain Tegmentum. 58
    1. Red Nucleus 58
    2. Nucleus Profundus Mesencephali and Substantia Nigra 59
    3. Interpeduncular Nucleus 62

Discussion 63

Nucleus of the Dorsal Supraoptic Decussation 63

Entopeduncular Nucleus. 64

Red Nucleus 65

Nucleus Profundus Mesencephali: Substantia Nigra 65

Summary 66

Bibliography 67

Index 69

Part I. Striatal Structures in Reptiles and Birds

Introduction

Comparative neurological studies of the evolutionary development of structures within the central nervous system of vertebrates have depended to a large extent upon morphological rather than functional criteria. Classical comparative anatomical studies, which have attempted to demonstrate homologies between parts of the brain in representatives of different vertebrate classes may be grouped under three general headings: 1. comparison of the embryological development of brain structures; 2. comparison in adult forms of the topographical relations of neuron groupings and fiber tracts, and of the morphology of cell types (cytoarchitectonics); and 3. analysis and comparison of fiber connections between particular cell groupings or regions. Of these three, the third encompasses functional relationships most directly, but even in well-defined fiber tracts the direction of conduction often remains indefinite, and the extent and activity of more diffuse systems is poorly known.

In recent years a number of investigations applying electrophysiological and degeneration methods to submammalian forms have been reported. Those most pertinent to the present studies include the papers of ARMSTRONG et al. (1953), KRUGER and associates (e.g. HERIC and KRUGER, 1966; KRUGER and BERKOWITZ, 1960; POWELL and KRUGER, 1960), GUSEL'NIKOV and SUPIN (1964) and KARAMYAN and BELEKHOVA (1964) on various reptiles, and of POWELL and COWAN (1961), KARTEN and REVZIN (1966) and REVZIN and KARTEN (1967) on the pigeon. These investigations have demonstrated connections between certain parts of the brain in greater detail and with greater assurance of correct polarity than had earlier anatomical works. Certain earlier postulated homologies of reptilian or avian brain structures with mammalian ones, which were based on morphological observations, have been challenged with results obtained by some of these investigators. These will be discussed in appropriate relation to the author's observations. However, the experimental analysis of submammalian brains is in its initial stages, and few structures in the brain, in a very few species, have been subjected to inquiry.

Recent advances in neurochemistry, and more specifically, neurohistochemistry, offer possible tools in the determination of evolutionary and functional homologies within the central nervous systems of animals widely separated on the evolutionary scale. Considerable information exists pertaining both to the overall topographical and the more detailed distribution of a number of enzymes and other chemical constituents within cell bodies, neuropil and fiber tracts of the brain in a number of species of adult mammals, including man. This large literature is extensively reviewed in the recent works of ADAMS (1965) and FRIEDE (1966). However, few "chemoarchitectonic" studies on the brains of submammalian forms are available for comparison with those on mammals. SINDEN and

SCHARRER (1949) investigated patterns of distribution of four enzymes in parts of the pigeon brain; SCHARRER and SINDEN (1949) studied the detailed distribution of three enzymes in the optic tectum of the pigeon and SCHARRER (1955) compared some phosphomonoesterases in the cells of the mesencephalic Vth nucleus of selachian fishes, reptiles and mammals. SHEN et al. (1955) studied the distribution of cholinesterase in the frog brain. FLEISCHAUER and HORSTMANN (1957) attempted (unsuccessfully) to find a pattern of zinc distribution in brains of a fish, a frog and a turtle. Papers by MASAI (1961a, b) have described glycogen distribution in brains of fish, amphibians, reptiles and birds, and works by MASAI and associates (1961a, b; 1966) have dealt with selected enzymes in brains of fish, amphibians and birds. Most recently, BERTLER et al. (1964) and FUXE and LJUNGGREN (1965) have applied the fluorescence histochemical method of Falck for catecholamines to the pigeon brain, with important results. These will be discussed below and in Part II of this series.

The work reported here, and that to be described in subsequent sections of this series, represents an initial attempt to suggest answers to two interrelated questions:

1. Is the concentration (or activity) of chemical constituents, in structures which now are generally accepted as homologous, similar in animals of different vertebrate classes ?

2. Can detailed knowledge of the chemical makeup of brain structures aid neuroanatomists in the determination of evolutionary homologies; permit the acceptance or rejection of as yet indefinite or doubtful homologies, or perhaps suggest new alternatives ?

Increased detailed knowledge of the comparative biochemistry of central nervous structures will eventually aid in the understanding of physiological differences in evolutionarily homologous systems in the brains of animals which differ widely in their ecological adaptations. Conversely, such knowledge could provide clues to the functioning of analogous systems which may have developed from different ancestral materials (for example, the hyperstriatum of birds has been referred to as "vicarious cortex" — ARIENS KAPPERS et al., 1960, Vol.III). It is hoped that the observations presented in this series may prove useful to comparative neurology in this context.

Materials and Methods

1. Animals[1]

Turtles. Specimens of three species were used; Slider Turtle (*Pseudomys* sp.) — one adult; Eastern Painted Turtle (*Chrysemys picta picta*) — four adults; Western Painted Turtle (*Chrysemys picta bellii*) five adults, two juveniles; Grecian Tortoise (*Testudo graeca*) — one adult.

Lizards. Specimens of two species were used: Carolina Anole (*Anolis carolinensis*) — seven adults; Texas Horned Lizard (*Phrynosoma cornutum*) — six adults.

Caimans. All specimens were South American Spectacled Caimans (*Caiman sclerops*) — four juveniles, none exceeding one foot in length.

Birds. Australian Shell Parakeets (*Melopsittacus undulatus*) — two young adults.

[1] I wish to thank Dr. RICHARD G. ZWEIFEL, Department of Herpetology, American Museum of Natural History, New York, for his examination and classification of the reptilian specimens.

Mice. These were domesticated laboratory mice (*Mus musculus*) either of the Swiss strain, or derivatives of special crosses[2] — 17 adults.

In all groups animals of both sexes were used, but as no sex differences were noted, the numbers of each sex are omitted. Apart from mice, animals were not maintained in the laboratory for any length of time. Reptile and bird specimens were purchased as required from local petshops, with the exception of two Eastern Painted Turtles caught wild at Croton, New York. When retained for more than a day or two, lizards were fed mealworms, and turtles and caimans were fed canned dog food or liver.

All of the varieties of animals chosen for this work were selected for their ready availability. Apart from this factor, the mouse, as a representative of the mammals, was selected because the small size of its brain made possible the production of complete sections, so that all parts of the brain could be surveyed in minimal time. The turtles were chosen for the large size of their brains, in comparison with those of the readily available small lizards, and because turtles remain nearer to the primitive reptilian stock, from which the mammals also were derived, than do the lizards and Crocodilians. The latter are more specialized groups of reptiles; the brain of the alligator shows great similarities to that of birds (PAPEZ, 1929).

2. Methods

a) Histochemical Preparations. The enzyme histochemical methods used in this study were chosen for their general acceptance and relative simplicity, and because all had been used extensively in published studies on mammalian brains. Whole or partial brains were sectioned transversely, horizontally or sagittally. The latter two planes were found to be most helpful in orientation and identification of structures in the reptilian brains.

Animals were killed by decapitation and the brain removed as rapidly as possible. For the preparation of fresh frozen sections, entire brains or parts thereof were immediately frozen over solid CO_2 or in the cryostat. The frozen brain specimens were sectioned in the cryostat at 12—16 μ and the sections mounted directly on cover slips. For the preparation of fixed frozen sections, entire or subdivided brains were fixed in cold formol-calcium overnight. Sections of 20 μ thickness were then cut on a freezing microtome and collected in cold distilled water. These sections were mounted on coverslips and air-dried before incubation. Two or more enzyme histochemical methods were applied to most specimens, usually on adjacent sections. When horizontal or sagittal series were produced of the small lizard brains, all sections were retained and three different methods applied to sets of adjacent sections. In series of the larger brains of turtles, alligators, parakeets and mice, the omission of several sections generally was alternated with the retention of sets of up to four adjacent sections, to each of which a different method was then applied. Portions of some of the larger brains also were processed without omission of sections, and several procedures carried out on adjacent sections.

Control sections were incubated without substrate. The histochemical methods used were as follows[3]:

1. Succinate dehydrogenase (SDH)[4]: modification of method of NACHLAS *et al.* (as given by BARKA and ANDERSON, 1965, p. 313). No prefixation. Incubations from 30 to 120 minutes.

2. Reduced nicotinamide-adenine dinucleotide tetrazolium reductase (NADH diaphorase): method of DEANE *et al.* (1962). No prefixation. Incubation 30 minutes.

3. Acid phosphatase (AcPase), fixed frozen sections: modification of the azo dye method of BURSTONE (PEARSE, 1960), or the metal salt method of GOMORI (1952). Incubations 60—90 minutes. These methods were applied only to brain specimens of mouse and *Phrynosoma*.

---

[2] The latter mice were kindly supplied by Dr. L. J. SMITH, Department of Anatomy, Albert Einstein College of Medicine.

[3] Nitro-BT and NADH (DPNH) were obtained from Sigma Chemical Company. Fast Red Violet LB salt (Burstone acid phosphatase method) was kindly donated by Verona Dyestuffs, Union, New Jersey.

[4] *Abbreviations.* The enzymes demonstrated by the histochemical methods will be abbreviated throughout this report as follows: SDH = Succinate dehydrogenase, NAD diaphorase = Reduced nicotinamide-adenine dinucleotide tetrazolium reductase, AcPase = Acid phosphatase, TPPase = Thiamine pyrophosphatase.

4. Acid phosphatase (AcPase), cryostat sections: modification of Gomori's (1952) metal salt method. Prefixed 60 minutes in cold buffered neutral formalin. Incubation 60 to 150 minutes.

5. Thiamine pyrophosphatase (TPPase): method of Novikoff and Goldfischer (1961). Prefixed 60 minutes in cold formol-calcium. Incubation 60 to 180 minutes.

b) Terminology: Identification of Structures. In the identification of structures in the brains of reptiles and birds, a great many papers were consulted, most of which will not be cited specifically in this section. The primary source of information, further references and nomenclature was, of course, the monograph of Ariens Kappers et al. (1960). The most frequently studied works on the turtle brain were those of Johnston (1915, 1923) and Papez (1935), and on the crocodilian brain those of Crosby (1917) and Huber and Crosby (1926). Papers by Shanklin (1930) on the brain of *Chameleon*, by Goldby (1934) on *Lacerta*, and by Cairney (1926) and Durward (1930) on *Sphenodon*, aided in the identification of structures in lizard brains. Huber and Crosby's (1929) study was extensively consulted in relation to the anatomy of the bird brain, and their terminology is used.

The identification of structures in the mouse brain, and the nomenclature used, was largely based on the atlases of König and Klippel (1963) and Meessen and Olszewski (1949), on the rat and rabbit respectively. Apart from these purely anatomical works, the histochemical atlases of Friede (1959a, b; 1961a, b), on the guinea pig and cat brain, were invaluable in the identification of structures in similar preparations of the mouse brain.

Observations and Comparisons

In the presentation of histochemical observations on the various submammalian brains studied, each structure, as it is taken up, will be introduced briefly from the literature most pertinent to its further description and interpretation. It is felt that this procedure is desirable in view of the widely varying terminology applied to submammalian brain structures in earlier anatomical descriptions.

In discussing each structure, histochemical obsevations first will be presented on the turtle brain, as representing the least specialized of the reptiles included. This will be followed as required by a brief discussion of possible mammalian homologies, and then by histochemical observations on the appropriate parts of the mouse brain and comparisons with the turtle. Obervations on corresponding structures in the brains of the two lizards and of the caiman will then be presented in the sequence most suitable for their best comparison with those in the turtle and mouse. The parakeet brain will be taken up last, as it is the most divergent from the primitive reptilian type, and most similar to that of the alligator.

Because of their complexity, and the conflicting analyses to be found in the literature on reptile brains, the amygdaloid nuclei will not be included in this report. Structures of the archistriatum will be described and discussed in a separate communication.

Palaeostriatum

1. Turtle

Johnston (1915), in his original analysis of the forebrain of turtles, described a "caudate" and a "lentiform" nucleus and a "dorsal ventricular ridge". From embryological and other observations he later (1916) concluded that the main part of the dorsal ventricular ridge originated as an infolding of the dorsal pallium medial to the piriform lobe, at the rhinal fissure, but that the caudo-ventral part of this ridge originated separately from an infolding of the piriform lobe itself, at the endorhinal fissure. The latter was concluded to be a part of the amygdaloid complex. JOHNSTON then reserved the term dorsal ventricular ridge for the antero-dorsal ridge, and referred to the caudo-ventral ridge (these are separated by a very shallow sulcus in turtle brains) as the amygdaloid ridge. This more restrictive terminology for the two structures will be followed here, and in future communications, in discussing the turtle brain. The term "dorsal ventricular ridge" will be abbreviated to "DVR" in the present report.

In a later report, JOHNSTON (1923) homologized the DVR with the major part of the caudate nucleus of mammals and his earlier named "caudate nucleus" was re-designated as the bed of the stria terminalis. The latter he believed to include only that part of the head of the caudate nucleus in mammals which is closely associated with the nucleus accumbens. His original lentiform nucleus he then subdivided into putamen and globus pallidus, by the recognition of a distinct large-celled entity within the nucleus which he believed to deserve the latter name. JOHNSTON believed the putamen, globus pallidus and "stria bed" to be evolutionarily the oldest representatives of the basal ganglia and thus included under the term paleostriatum. He reserved the term neostriatum for the main part of the caudate nucleus.

In acid phosphatase preparations of turtle brains, JOHNSTON's putamen was seen only as an area containing chiefly small weakly stained cells, which was not particularly noticeable (Fig. 4). But in preparations incubated for SDH and NAD diaphorase, intense staining of the neuropil caused this nucleus to stand out vividly (Figs. 3, 6, 11). This was especially striking for SDH. Horizontal sections revealed the putamen to be a somewhat sausage-shaped structure, extending along the lateral surface of the hemisphere in a gentle curve. From the rostral tip of the hemisphere (where it became indistinguishable in the oxidative enzyme preparations from JOHNSTON's anterior olfactory nucleus), it extended to a point about two-thirds of the distance to the caudal pole (Fig. 11). (Unlike the putamen of mammals, that of the turtle, as identified by JOHNSTON, is almost bare of cortex laterally. It is covered, according to JOHNSTON, only by a thin layer of small cells belonging to the nucleus of the lateral olfactory tract.) The neuropil of the more caudal part of the putamen stained somewhat more heavily for SDH and NAD diaphorase than the rostral part, and AcPase incubation revealed rather more strongly stained cells and a more clumped cell distribution caudally. In standard incubations for SDH and NAD diaphorase, the neuropil was generally dark enough to obscure cell body staining, although in the more caudal area some still darker cells could be discerned. With shorter incubations for SDH, darkened cell somas could be made out in all parts. SDH staining, in particular, also delineated a great many cell processes in this area.

The globus pallidus of the turtle consists of a group of large cells embedded in the fan-like portion of the lateral forebrain bundle as it spreads out adjacent to the putamen. As in mammals, the cells are large and multipolar. Histochemical studies showed these cells and their processes to be intensely reactive for SDH (Figs. 5, 6, 12), NAD diaphorase (Fig. 15), and AcPase (Fig. 4). The neuropil surrounding the cells was lightly stained and did not obscure the perikarya in the dehydrogenase preparations.

2. Mouse

Histochemical studies on the mouse brain showed that the globus pallidus and putamen of the turtle and mouse differed little with respect to the relative activities of the enzymes mentioned. The neuropil of the fused putamen-caudate of the mouse was intensely stained for the dehydrogenases (Figs. 40, 41) and the small cells were weakly stained for AcPase. Cells of the globus pallidus in the mouse were strongly stained for SDH, NAD diaphorase, and AcPase. There is a more distinctly differentiated neuropil associated with the globus pallidus neurons in the mouse than in the turtle, and it was weakly stained for the dehydrogenases (Fig. 39). These observations on the mouse are similar to those made previously by Ortmann (1961) and Shimizu and Morikawa (1957) and to those made on other mammals by Friede (1961a, guinea pig), Friede and Fleming (1962, man; 1963, monkey), Shimizu and Morikawa (1957, rat, rabbit, guinea pig) and Shimizu et al. (1957, rabbit). In the guinea pig, because of its heavy staining, the putamen was used as a standard for comparison of densitometric measurements of SDH by Friede (1961a).

3. Lizards

In lizards, a globus pallidus has not been specifically identified by name. Durward's (1930) figures showed large cells within the lateral forebrain bundle in the hemisphere of Sphenodon, but he did not label them. However, in his text (p. 23), he implied a relationship of these to Johnston's "lateral large celled lenticular nucleus." More recently, Armstrong et al. (1953), in Anolis, compared the large motor-type cells of their ventrolateral palaeostriatum with the cells of the globus pallidus in mammals.

The present studies indicated the globus pallidus of lizards to be histochemically similar to that of the turtle. In both Anolis and Phrynosoma, the perikarya of large cells associated with the lateral forebrain bundle in the telencephalon stained intensely for SDH (and for NAD diaphorase in Anolis). In both lizards, the long processes of these multipolar cells also stained prominently for SDH (Figs. 26, 29). In neither species of lizard were the globus pallidus neurons as prominent in AcPase preparations as those of the turtle or mouse, as many of the cells were not strongly stained. In Anolis, TPPase staining was strong in many of these cells and their processes.

Characterization of a putamen (according to Johnston's nomenclature) in lizard brains, by examination of coronally sectioned histochemical material seemed at first a problem. It had not been identified as such in the literature, nor did published figures of reptile brains other than Johnston's of the turtle seem to outline clearly a suitable cell mass in the area surrounding and rostral to the lateral forebrain bundle. This general area has been referred to as "somatic striatum" by several authors (e.g. Cairney, 1926; Durward, 1930) and as palaeostriatum by others (e.g. Shanklin, 1930; Goldby, 1934). However, in horizontally sectioned Anolis brain, SDH incubation was found to delineate a rounded mass of fairly deeply stained neuropil, which formed a cap around the anterior pole of the lateral forebrain bundle and enclosed the globus pallidus (Figs. 28, 29). In adjacent sections incubated for NAD diaphorase, this neuropil was less prominently stained. In other adjacent sections, AcPase staining revealed a corresponding area chiefly made up of small, rather weakly stained cells. This area was bounded rostrally by a curved margin formed by the large, heavily granulated cells of the rostral cortex. With the aid of these horizontal sections, and of certain observations made on the caiman brain, an area of fairly strong SDH staining corresponding in position to the putamen of the turtle could be outlined in transverse sections of the lizard brains (Figs. 25, 26). This structure in the lizard brain thus differed from the putamen of turtles and from the caudate-putamen of mammals in the considerably lesser oxidative enzyme activity of its neuropil. As in the turtle and mouse, scattered larger cells with high SDH activity were observed in this part of the brain of *Anolis*.

4. Caiman

CROSBY's (1917) account of the forebrain of the alligator made use of a different set of terms than used by others, but she compared many of these with JOHN-STON's (1915) nomenclature. She considered that: 1, her "dorso-lateral area" was at least partly comparable to JOHNSTON's dorsal ventricular ridge; 2. her "ventro-lateral large-celled area" was similar to JOHNSTON's lentiform nucleus; 3. her "ventro-lateral small-celled area" was equivalent to JOHNSTON's nucleus caudatus (which is JOHNSTON's (1923) stria bed). Her intermedio-lateral area she believed to become continuous with the dorso-lateral area caudally. ARIENS KAPPERS *et al.* (1960, Vol. III, p. 1322) believed the intermediolateral area to represent part at least of JOHNSTON's putamen. CROSBY observed that the anterior part of the dorsolateral area received almost exclusively somatic sensory fibers, from the thalamus, whereas caudal parts of this area, especially lateral and dorsolateral, received largely olfactory fibers.

Histochemical preparations of the caiman forebrain showed similarities to that of both lizards and turtle. Comparison with the turtle brain aided materially in the interpretation of the observations. Coronal sections incubated for SDH clearly showed large intensely stained globus pallidus neurons among the fanned-out fibers of the lateral forebrain bundle (Figs. 16, 22, 23). Their appearance was similar to that in the other reptiles, although they were relatively more numerous than in the other species. They did not stain very strongly for AcPase, unlike those of turtles, but similar to those of the lizards. Their spatial relations were particularly well displayed in certain sagittal sections which showed the greater extent of the ventral peduncle of the lateral forebrain bundle in its passage between the hemisphere and the ventral thalamus (Fig. 31).

In coronal sections, the rostral fanned-out extremity of the lateral forebrain bundle was observed to be capped by a crescentic area of neuropil which stained more deeply for SDH than adjacent dorsal areas, although still to only a moderate degree (Fig. 16). It contained some scattered large deeply stained cells superficially resembling those of the globus pallidus. The AcPase reaction in adjacent sections showed this area to be made up of a mass of small, rather pale cells, separated slightly from much larger and more deeply stained neurons dorsal and lateral to it. Sagittal sections revealed it to be a rounded mass with rather definite boundaries rostrally and caudally, as well as dorsally (Figs. 20, 31). Although it differed in the comparative pallor of its SDH staining, it was similar in relative size and position to JOHNSTON's putamen in the turtle, as seen in comparable sagittal sections (Figs. 30, 31). There seemed to be little doubt, that this structure in the caiman brain was homologous with Johnston's putamen in the turtle. It then was notable for its low SDH activity in comparison with its homologue in both turtle and lizards, as well as with the putamen-caudate complex of the mouse and other mammals.

The "putamen" of the caiman, as outlined here, does not include Crosby's intermediolateral area (suggested by Ariens Kappers to represent part at least of the putamen of mammals), but lies entirely ventral to that cell mass. Nor does it include the large neurons forming its dorsal and lateral boundaries, and which apparently represent Crosby's ventrolateral large-celled area. Therefore, it can only be represented by a part of Crosby's ventrolateral small-celled area.

5. Parakeet

Although a terminology different from that used for either reptile or mammal brains has been applied in the description of many parts of the brains of birds, the equivalence of a number of structures in the bird brain to reptilian or mammalian counterparts has been demonstrated. The avian palaeostriatum primitivum has long been considered to be an homologue of the mammalian globus pallidus (Huber and Crosby, 1929). This interpretation has been recently supported by comparative electron microscopic investigations, with the chicken as the avian representative (Fox et al., 1966). In the interest of unity in terminology, the name globus pallidus will therefore be used here, in the description of this structure in the parakeet brain.

The palaeostriatum augmentatum (also called mesostriatum), which surrounds, or forms a cap over the globus pallidus, has been thought to be homologous with the rostro-ventral part of the caudate nucleus of mammals (Crosby et al., 1966), but also believed to include parts of both the caudate nucleus and putamen of mammals (Ariens Kappers et al., 1960). Huber and Crosby (1929) likened this part of the avian brain to the ventrolateral large-celled area in the brain of the alligator.

In the histochemical preparations of the parakeet brain, the globus pallidus and palaeostriatum augmentatum bore a close resemblance to the globus pallidus and Johnston's putamen in the turtle, and the corresponding parts of the brain in the other reptiles. Neurons of the globus pallidus, like those of the reptiles and of mammals, were strongly active for SDH and NAD diaphorase (Figs. 34, 36). Like those of the lizards and alligator they were not strongly stained in AcPase preparations. The neuropil of the palaeostriatum augmentatum was intensely active for SDH, especially at its more dense rostral end, where fewer fiber bundles passed through it (Fig. 38). This mass of neuropil, like the putamen-caudate of the mouse, and the putamen of the turtle, was the most heavily stained structure in the telencephalon for SDH. It was approached in activity only by the hyperstriatum accessorium and ectostriatal area (see below). AcPase incubations revealed a predominance of small and weakly stained neurons in the palaeostriatum augmentatum, but a scattering of large highly active neurons was present (Fig. 37). The latter seemed to be more prevalent than those in the putamen of the turtle, alligator or mouse.

Dorsal Striatum

1. Turtle

As mentioned above, JOHNSTON (1923) believed that his dorsal ventricular ridge (DVR) was homologous with the main part of the caudate nucleus of mammals. A differentiated part of the DVR, although not labelled in JOHNSTON's (1915) drawings, was described in his text as a "somewhat quadrilateral area" with "almost the appearance of an independent cell mass". This he called the "core-nucleus" of the DVR.

In coronal sections of turtle brains incubated for SDH or NAD diaphorase' the core nucleus stood out as a heavily stained block of neuropil. It was somewhat less strongly stained for SDH than JOHNSTON's putamen, which adjoined its ventral aspect (Fig. 7), but was more active for NAD diaphorase than the latter. Its box-like shape could be appreciated by comparison of sections in the three planes. This core nucleus, in *Chrysemys*, was found by CRAIGIE (1941) to rank second only to the cochlear nucleus in richness of the capillary bed. The turtle putamen (lentiform nucleus) also was found by CRAIGIE to be notably well vascularized. Thus the histochemical observations on SDH activity in the neuropil correlated in a general way with vascularity, although the ranking order of the two areas was reversed.

Besides being strongly stained for the dehydrogenases tested, the core nucleus was "browned" in the AcPase and TPPase reactions. It contained many fibrils prominently delineated by most of the enzyme methods used, as well as by gallocyanin staining. Many of these fibrils were oriented in the direction of JOHNSTON's putamen. The core nucleus appeared to contain mainly large neurons, whose perikarya were strongly stained for SDH, NAD diaphorase, AcPase and TPPase (Figs. 8, 10, 14).

The neuropil of the remaining (antero-medial) portion of the DVR was more weakly stained for SDH than that of the core nucleus (Figs. 7, 9). NAD diaphorase was fairly strong in its ventral part and weaker dorsally. No part showed "browning" in AcPase and other lead sulfide preparations, and no fibrils could be seen. The neuron bodies in the antero-medial DVR were strongly stained both for SDH and NAD diaphorase. These cells, the largest of which are distributed in a "cortical" (subventricular) arrangement of small clumps, were among the most prominently stained neurons for AcPase in the turtle brain (Figs. 4, 8, 10).

Thus, it became apparent that the DVR of the turtle was clearly divided, not only cytologically, but histochemically into two parts with quite different characteristics. Some possible implications of this will be discussed further in relation to observations on the other reptiles and bird.

2. Mouse

In the mouse, the putamen and caudate nucleus are fused and indistinguishable, and collectively they contain mainly small cells, with a scattering of larger ones. As already described for the putamen, the neuropil of this complex was intensely stained for SDH and NAD diaphorase. No area could be found in which cells were especially prevalent which stained sufficiently strongly to be distinguishable over the neuropil for these enzymes. This is to be contrasted to the histochemical

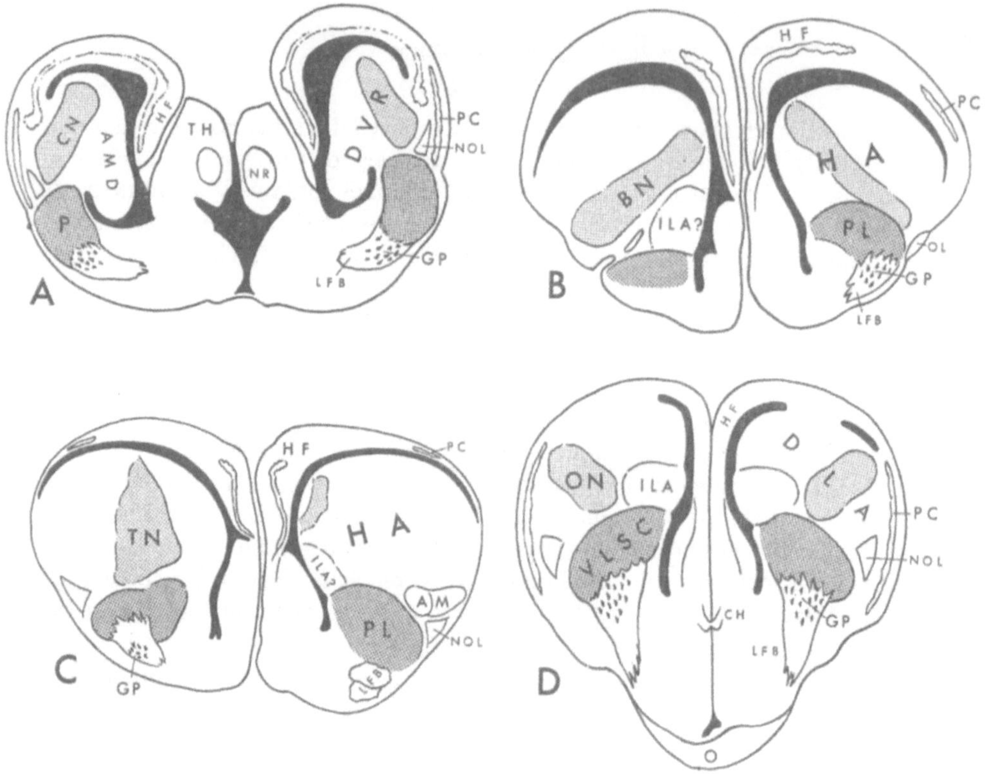

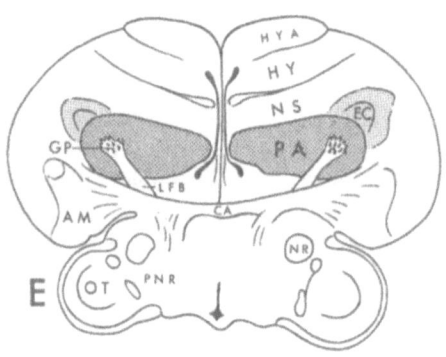

Fig. 1 A—E. Semi-diagramatic projection drawings of coronal sections through the rostral telencephalon of the five submammalian vertebrate types represented in this report. The original sections were incubated to demonstrate succinate dehydrogenase activity, and the structures are drawn as revealed by this treatment. The figures are drawn to approximately uniform size, rather than to uniform magnification. Ventricular spaces are filled in solid black. In each drawing, the labelling represents largely the terminology applied in a classical anatomical description of that type — for which the reference is given in each case. Attention is drawn to similarities in the relative positions of the areas labelled *CN* (A), *BN* (B), *TN* (C), *ON* (D) and *EC* (E), especially with respect to the areas labelled *P* (A), *PL* (B and C), *VLSC* (D) and *PA* (E). The spatial relation of the globus pallidus to each of the latter group of structures is similar. The areas labelled *ILA* and *ILA?* in B, C and D may be comparable to that labelled *AMD* in A. Further similarities in topographical relations, and in the histochemical acitivity and structure of the striatal components illustrated in each animal type are described in the text, and photographically illustrated in the plate figures. A. Turtle (*Pseudomys*) JOHNSTON (1923). B Lizard (*Phrynosoma*) GOLDBY (1934). In this diagram, the right-hand half of the brain is drawn from a section taken somewhat more caudally than that of the left-hand half, in order to illustrate well both the "band nucleus" and the globus pallidus and lateral palaeostriatum. C Lizard (*Anolis*) GOLDBY (1934). In this diagram also, the right-hand half of the brain is drawn from a section taken slightly more caudally than that of the left-hand half, to illustrate well both the "triangular nucleus" and the area (*ILA*?) thought to correspond to the intermediolateral area of the alligator. D Caiman (*Caiman*) CROSBY (1917). E Parakeet (*Melopsittacus*) HUBER and CROSBY (1929)

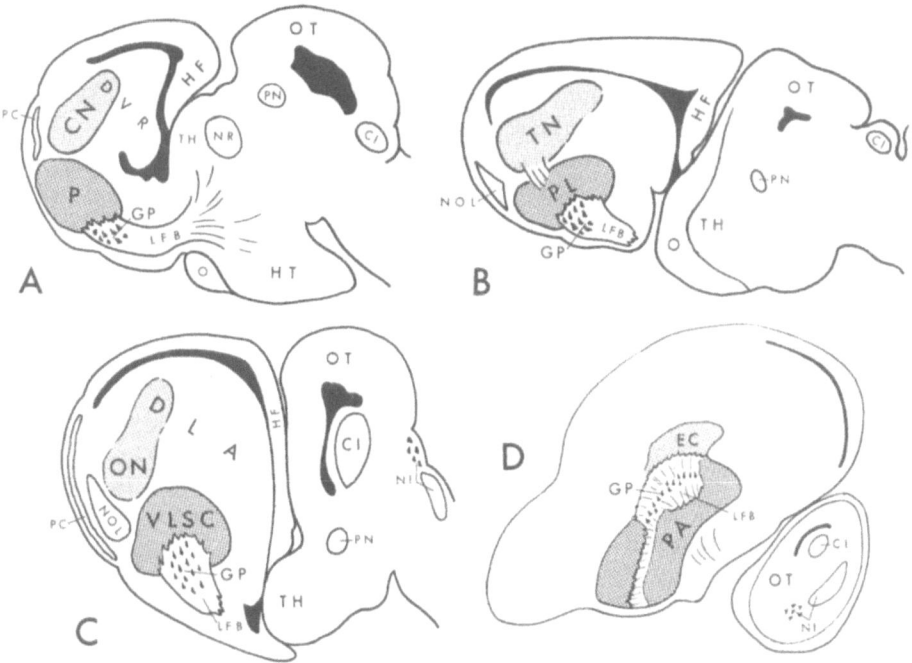

Fig. 2 A—D. Semi-diagramatic projection drawings of sagittal sections through the fore- and midbrain regions of four of the submammalian vertebrate types represented in this report. As in Fig. 1, the sections from which these drawings were made were incubated for the demonstration of succinic dehydrogenase activity. Reference should be made to the legend of Fig. 1 for explanation of the terminologies employed, and the significant comparisons to be made between the different species shown. The ventricular spaces are filled in solid black. A Turtle (*Chrysemys*), B Lizard (*Anolis*), C Caiman (*Caiman*), D Parakeet (*Melopsittacus*)

Abbreviations for Figures[1]

AM	amygdaloid structures	*IP*	interpeduncular nucleus
AMD	antero-medial part of dorsal ventricular ridge	*IS*	nucleus isthmi
		LFB	lateral forebrain bundle
BN	"band nucleus"	*NOL*	nucleus of the lateral olfactory tract
BO	nucleus of the basal optic root	*NR*	nucleus rotundus
C	cerebellum	*NS*	neostriatum
CA	anterior commissure	*O*	optic tract
CH	hippocampal commissure	*OC*	optic chiasma
CI	inferior colliculus	*OL*	lateral olfactory tract
CN	core nucleus of the dorsal ventricular ridge	*ON*	"oval nucleus"
		OT	optic tectum
DLA	dorso-lateral area	*P*	putamen
DVR	dorsal ventricular ridge	*PA*	palaeostriatum augmentatum
EC	ectostriatum	*PC*	piriform cortex
GP	globus pallidus	*PL*	lateral palaeostriatum
H	habenula	*PN*	pretectal nucleus
HA	hypopallium anterior	*PNR*	nuclei of pretectal region
HF	hippocampal formation	*RE*	nucleus reuniens
HT	hypothalamus	*TH*	thalamus
HY	hyperstriatum	*TN*	"triangular nucleus"
HYA	hyperstriatum accessorium	*VLSC*	ventro-lateral small-celled area
ILA	intermedio-lateral area	*III*	oculomotor nucleus
ILA ?	presumptive intermedio-lateral area	*IV*	trochlear nucleus

[1] In all figures, representations of sagittal sections are oriented so that the rostral pole lies to the left. Horizontal sections are oriented so that the rostral pole lies at the top of the figure.

observations on the DVR complex of the turtle. Friede (1961) observed that the scattering of large neurons in the separate putamen and caudate nucleus of the guinea pig were barely discernible against the heavy staining for SDH in the two neuropils. In a mapping of NAD diaphorase distribution in the human brain, Friede and Fleming (1962) gave similar densitometric measurements for staining intensity in the putamen and caudate nucleus, and under a common heading, mentioned that large cells in the putamen occasionally showed a stronger reaction. Such observations imply the existence of little or no histochemical difference between the two areas in mammals.

The core nucleus of the DVR in the turtle resembled the caudate nucleus of mammals in the strong oxidative enzyme activity in its neuropil and in the perikarya of the larger neurons. However, it was quite different from the mammalian caudate nucleus in containing *preponderently* large sized neurons, with strong AcPase activity. The latter attributes differentiated it distinctly from Johnston's putamen in the same individual, whereas in mammals the caudate nucleus and putamen are essentially identical, even when anatomically separated.

The remaining portion of the DVR in the turtle shared few histochemical or cytological features with either the putamen of the turtle or the putamen and caudate nucleus of mammals. These observations could be interpreted to suggest that if any part of the DVR is an homologue of the caudate nucleus of mammals such homology might be restricted to the core nucleus, rather than encompassing the whole of the structure.

3. Caiman

In both lizards and caiman, preparations incubated for SDH revealed areas of extremely intense neuropil staining in the rostral part of Crosby's dorsolateral area (DVR of Johnston). However, since the interpretation of this part of the lizard brain was materially aided by the observations made on the caiman, these will be described first in this section.

Coronal sections of the caiman forebrain, incubated for SDH and NAD diaphorase, showed heavily stained areas of a roughly oval shape lying in the lateral part of the dorsolateral area (Figs. 16, 17). These were positioned above the putamen (Johnston's usage), and had similar spatial relations to the nucleus of the lateral olfactory tract, piriform cortex and "putamen" as did the core nucleus of the DVR in the turtle (Fig. 1). They did not show a sharp margin, but the staining intensity graded off rather steeply in all directions. Although this neuropil was stained much more deeply than that of the core nucleus of the turtle, like the latter this area could be seen to contain many large deeply stained neurons. Cell bodies scattered throughout the dorsolateral area were positive for SDH, but those within the "oval nucleus" were most strongly so. AcPase staining was strong in many neurons of the caiman dorsolateral area, but those in the "oval nucleus" and the area lateral to it were especially large and dark, and rather widely spaced. AcPase incubation also led to a diffuse "browning" in the "oval nucleus".

Examination of sagittal sections incubated for SDH showed the intensely stained "oval nucleus" to be a somewhat elongate structure, meeting the putamen at its rostral end, and extending caudalward and upward toward the floor of the lateral ventricle (Figs. 20, 31). Its caudalward extent was about half of the total length of the dorsolateral area. With AcPase staining, the background "browning" and large dark cells again were evident (Fig. 21). These sagittal AcPase preparations also emphasized the distinct differentiation of the caudal dorsolateral area, with its large, widely spaced, deeply stained neurons and pale ground color, from the ventral and rostral parts. Comparison of sagittal sections of the brains of turtle and caiman showed that the spatial relationships between the "putamens" of the two animals, and the core nucleus in the turtle, or the "oval nucleus" in the caiman, were entirely similar (compare Figs. 30 and 31). Thus, in spatial relations, cytology and enzymatic activities, the core nucleus and the "oval nucleus" in the turtle and alligator appeared to be homologous.

The intermediolateral area of the caiman, which appeared in rostral coronal sections, was conspicuous in AcPase preparations for its large deeply stained neurons (Fig. 24). It could be seen to be separated from the "putamen" at the ventricular margin by a small sulcus, the fissure neopalaeostriatica (HUBER and CROSBY, 1929). In reminiscence of the DVR of the turtle, it formed a distinct, although much smaller bulge into the ventricle. This was most pronounced at the rostral tip of the structure. As CROSBY noted, the intermediolateral area appeared to be continuous in histochemical sections with the main large-celled mass of the dorsolateral area. Like the latter, its neuropil was relatively weakly stained for SDH, and its neurons strongly so. CROSBY's suggestion that it probably is part of the dorsolateral area was supported by the similarity in enzymic activity.

Above the intermediolateral area, and dorso-medial to the "oval nucleus" neurons in AcPase preparations were more lightly stained and smaller than those below. This dorso-medial smaller-celled area persisted from the rostral end of the dorsolateral area almost to the caudal pole, becoming increasingly smaller caudally. (It was not delineated by CROSBY.) In coronal SDH preparations, it became bounded below by a "tail" of more deeply stained neuropil, which extended from the "oval nucleus" medially toward the wall of the lateral ventricle (Figs. 17, 18). In AcPase preparations, the cells in this "tail" appeared to be small and the neuropil was somewhat "browned". The appelation "tail" is actually a misnomer, for the persistence of these structures throughout most of the length of the dorso-lateral area in transverse series showed them to be actually plate-like in shape. These plates of tissue thus formed a boundary between a smaller-celled area above and a larger-celled main area below. The latter was continuous rostrally with the intermediolateral area as discussed above. These platiform structures revealed in the caiman brain by SDH staining will be of interest in comparison with similar appearing structures found in SDH preparations of the parakeet brain.

4. Lizards

Somewhat the same histochemical picture as in the caiman could be distinguished in the rostral dorsolateral area of *Anolis*. The most prominent difference was that transverse sections incubated for SDH or NAD diaphorase showed a sharp separation (almost as distinct as if cut off with a knife) between an intensely stained lateral area and a weakly active medial one. The heaviest staining in the lateral part seemed to be "piled up" at the boundary, and graded off in other

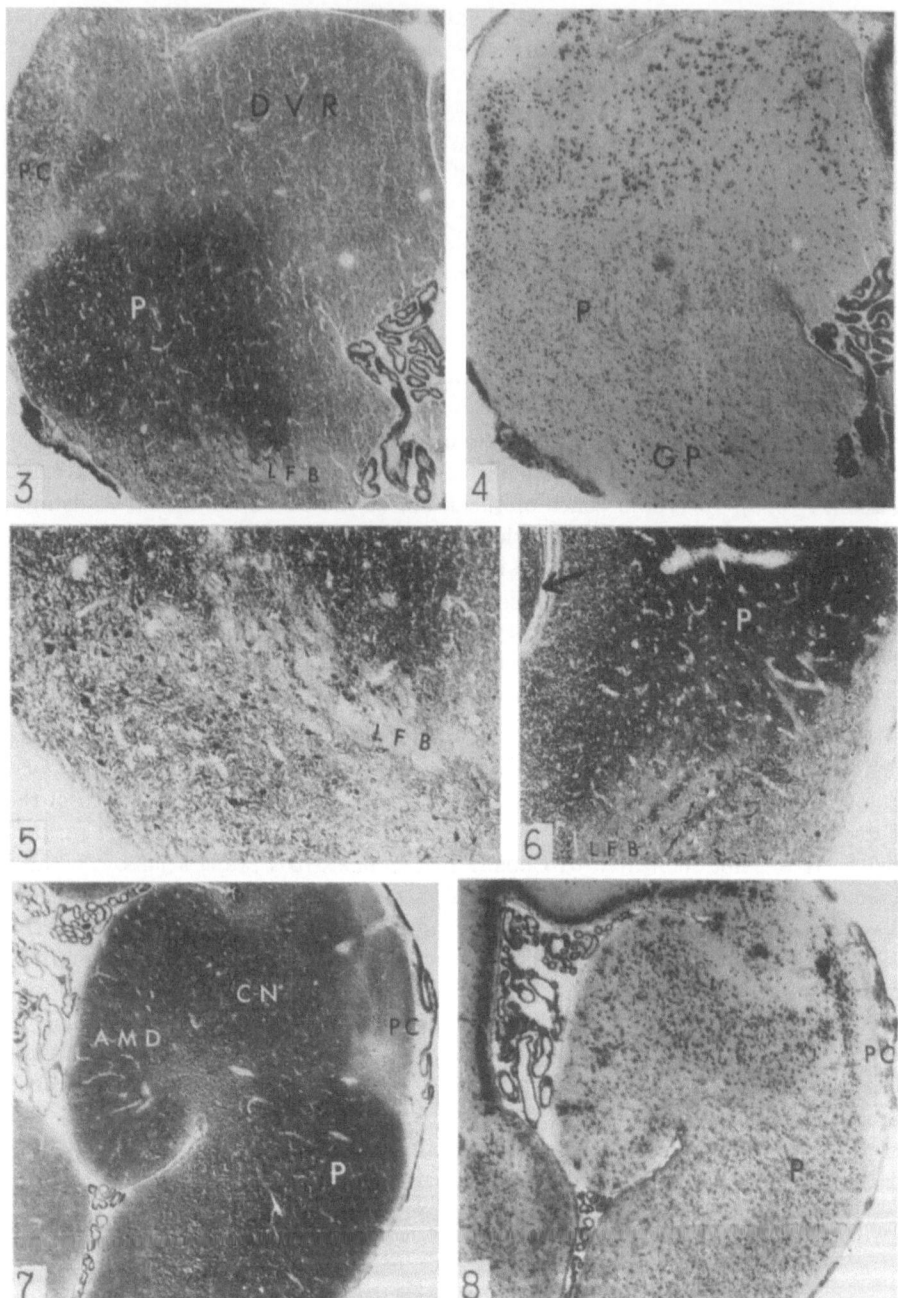

Fig. 3—8

Fig. 3. Western Painted Turtle. Sagittal section, SDH. Telencephalon. The deeply stained area represents Johnston's putamen, with strongly active neurons of the globus pallidus among fibers of the lateral forebrain bundle below it. The dorsal ventricular ridge, bulging into the lateral ventricle above the putamen, is weakly stained, as is piriform cortex, appearing at the left. Choroid plexus in the ventricle is highly active. ×30

directions (Figs. 25, 28). Since the most deeply stained part of the lateral area was roughly triangular in shape in transverse sections, it will be referred to in further reference as the "triangular nucleus" (Fig. 1). It contained scattered large intensely stained neurons, as seen in the "oval nucleus" of the caiman and the core nucleus of the turtle. The general size of the "triangular nucleus", and its orientation in relation to the "putamen" and other structures appeared similar to that of the above-mentioned areas in the caiman and turtle brains (again best demonstrated in sagittal sections — Figs. 30—32), although its position was closer to the midline than the caiman area. Horizontal sections showed that the weakly stained medial area extended laterally around the rostral tip of the triangular area (Fig. 28).

Acid phosphatase preparations showed weakly active smaller cells in the medial area and moderately strongly stained larger cells in the "triangular nucleus". Still larger and more heavily stained neurons were scattered throughout the latter area. In the *Anolis* brain (but not in the other species), AcPase incubation resulted in the staining of blood vessels, a circumstance which created annoying interference with the observation of neuronal enzyme activity. However, it did permit the observation that vascularization in the "triangular nucleus" appeared to be greater than in other adjacent parts of the striatum (Fig. 27). This parallels (CRAIGIE's, 1941) observation on the vascularity of the core nucleus in the turtle, cited above.

SDH-incubated transverse sections of a *Phrynosoma* brain also showed a deeply stained area with a fairly sharp medial boundary, although its shape and spatial orientation differed from the "triangular nucleus" in *Anolis*. In such

Fig. 4. Western Painted Turtle. Sagittal section, AcPase. Section adjacent to that in Fig. 3. Neurons of the globus pallidus show strong activity, but those of JOHNSTON's putamen are weakly stained. The large neurons of the dorsal ventricular ridge are highly active, as are cells of the general cortex and piriform cortex. × 30

Fig. 5. Western Painted Turtle. Sagittal section, SDH. Higher magnification of part of section shown in Fig. 3. Neurons of the globus pallidus and their processes show strong activity. Some strongly stained cells may be discerned within the deeply stained neuropil of the putamen. Fiber bundles of the lateral forebrain bundle are unstained. 80×

Fig. 6. Eastern Painted Turtle. Coronal section, SDH. Neurons and processes of the globus pallidus, appearing among the fibers of the lateral forebrain bundle, are strongly stained. Neuropil of JOHNSTON's putamen contains high enzymatic activity. The parts adjacent to the lateral ventricle (arrow) show weak activity. Blood vessels in the putamen appear as open strips. × 42.5

Fig. 7. Eastern Painted Turtle. Coronal section, SDH. Strong activity appears in the core nucleus of the dorsal ventricular ridge, although somewhat less than in JOHNSTON's putamen. The neuropil of the antero-medial part of the dorsal ventricular ridge is much less strongly stained. Piriform cortex is weakly active. × 20

Fig. 8. Eastern Painted Turtle. Coronal section, AcPase. Section adjacent to that in Fig. 7. Neurons of the core nucleus are strongly stained, while the smaller cells of the putamen are weakly so. In the antero-medial dorsal ventricular ridge, the large highly active neurons tend to form peripheral clumps. Hippocampal, dorsal cortical and piriform cortical neurons all show strong activity. Sinking in from the dorsal cortex, just above the core nucleus, are clumps of cells representing JOHNSTON's "pallial thickening". This structure is larger rostrally. × 20

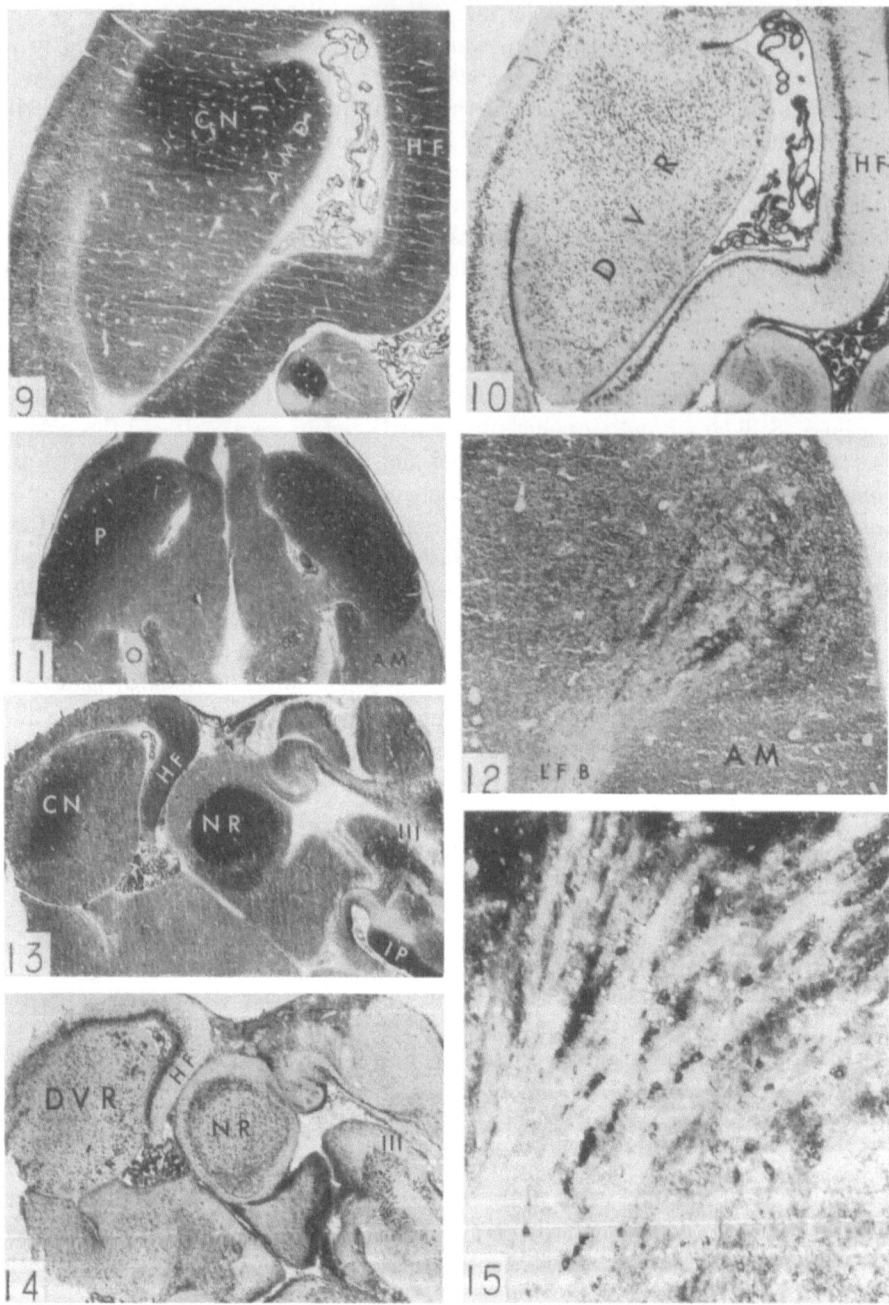

Fig. 9—15

Fig. 9. Grecian Tortoise. Horizontal section, SDH. Hemisphere. The dorsal ventricular ridge forms the large structure bulging into the lateral ventricle. The core nucleus shows as high or higher enzymatic activity as the hippocampal cortex. The antero-medial and caudal parts of the dorsal ventricular ridge show much less activity in the neuropil, although darkly stained neuron bodies may be seen. The piriform cortex is quite weakly stained. The dark structure at lower left is part of the habenular nucleus in the diencephalon. ×20

sections, it formed a wide band with its long axis oriented diagonally across the hemisphere, so that its medial end approached the dorso-medial intraventricular bulge of the dorsolateral area (Fig. 1). This positioning thus somewhat resembled that of the core nucleus in the turtle brain. Because of this orientation, the medial weakly active area apparently corresponding to that in *Anolis* lay ventrally in the dorsolateral area, rather than dorsally as in *Anolis*. In the "band nucleus", SDH activity was not only strong in the neuropil, but also was strong in the many large neuron bodies present in that structure.

Acid phosphatase incubation, as in *Anolis*, showed weak activity in the cells of the medial SDH-poor sector and moderate to strong activity in all cells of the lateral areas. AcPase staining was greatest in the neurons of the "band nucleus", and "browning" was present there.

In both *Anolis* and *Phrynosoma*, limited areas containing rather large cells which stained especially strongly for AcPase appeared in a ventral portion of the medial sector, just above the "putamen", and slightly rostral to the hippocampal commissure. These resembled CROSBY's intermediolateral area, but did not extend as far rostrally as in the caiman (Fig. 1). As mentioned above, CROSBY believed that this area in the alligator brain became continuous with the caudal dorso-lateral area, and probably was a part of it. These similar areas in the lizard brains also appeared to represent a part of SHANKLIN's (1930) pars medialis of the neo-striatum, which he believed was continuous caudally with his pars medialis of the archistriatum. (SHANKLIN used the term "intermediolateral area" for a cell

Fig. 10. Grecian Tortoise. Horizontal section, AcPase. Section adjacent to that in Fig. 9. Neurons of the core nucleus and of the peripherally-arranged clumps in the antero-medial and caudal dorsal ventricular ridge are strongly stained. Highest activity is seen in neurons of the hippocampal formation. The piriform cortical cells also show strong staining. $\times 20$

Fig. 11. Western Painted Turtle. Horizontal section, NAD diaphorase. Outstanding enzymatic activity is shown by the sausage-shaped putamen (of JOHNSON). It cannot be distinguished in such a preparation from the anterior olfactory nucleus, which protrudes at the rostral pole into the olfactory ventricle. The bulge into the lateral ventricle caudal to the putamen is formed at this level by the weakly active amygdaloid ridge. The optic tracts are unstained. $\times 8$

Fig. 12. Grecian Tortoise. Horizontal section, SDH. Globus pallidus neurons and processes, associated with the fanning-out of the lateral forebrain bundle in the telencephalon, are strongly active. This section is cut below the level of the putamen. The weakly stained area caudal to the globus pallidus includes various amygdaloid structures. $\times 45$

Fig. 13. Western Painted Turtle. Sagittal section, SDH. The three structures which show strongest enzymatic activity in this section are: the nucleus rotundus of the thalamus, the core nucleus of the dorsal ventricular ridge, and the hippocampal cortex. $\times 11.5$

Fig. 14. Western Painted Turtle. Sagittal section, AcPase. Section adjacent to that in Fig. 13. The especially strong activity for this enzyme shown by the clumped neurons of the dorsal ventricular ridge and those of the hippocampal formation is apparent. Strong activity is also shown by cells of the core nucleus, thalamic nuclei, including nucleus rotundus, and oculomotor-trochlear nuclei. $\times 11.5$

Fig. 15. Western Painted Turtle. Horizontal section, NAD diaphorase. Deeply stained neurons of the globus pallidus among the fibers of the lateral bundle as it fans out in the telencephalon. Deeply stained neuropil at top of picture is part of JOHNSTON's putamen. $\times 80$

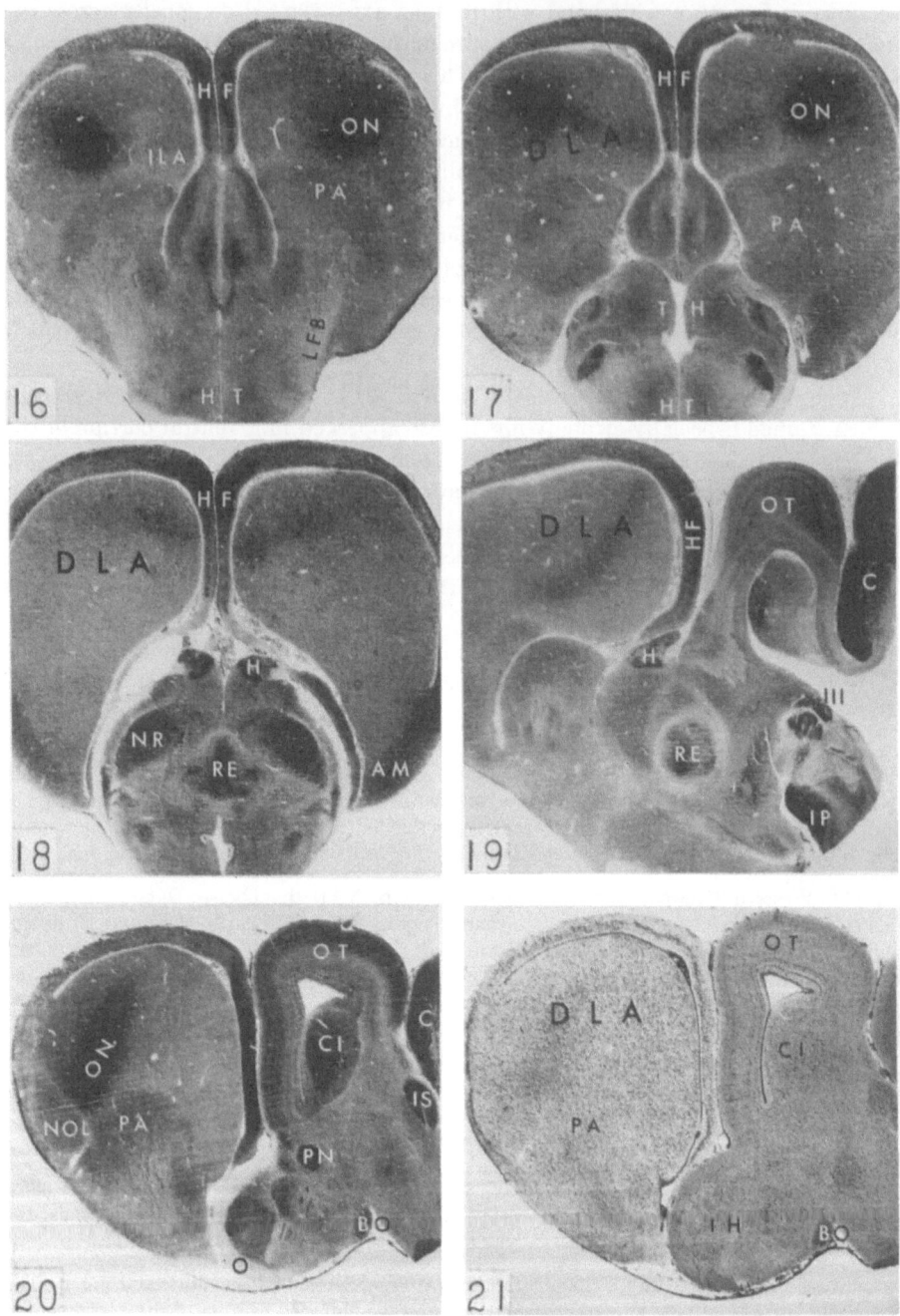

Fig. 16—21

Fig. 16—18. Caiman. Coronal sections, SDH. A series to illustrate the relations of the various paleostriatal and dorsal striatal structures considered in this report. In Fig. 16, the palaeo-striatum augmentum (or CROSBY'S ventro-lateral small-celled area) is seen as a crescentic area capping the fanning-out of the lateral forebrain bundle. It is separated from the dorsal striatum by a small ventricular sulcus. Its neuropil is more stained than most of the dorsal

(continued page 25)

grouping which appears from his figures to correspond with CROSBY's ventro-lateral large-celled area, rather than her intermediolateral area. This usage of the term appears to have been perpetuated by KRUGER and BERKOWITZ, 1960.) In the lizards, as well as in the caiman, activity for SDH in the neuropil of these medial areas was similar to, and continuous with, that of the caudal dorsolateral area.

5. Parakeet

In birds, the portion of the striatum overlying the palaeostriatum has been divided into a neostriatum, ectostriatum, hyperstriatum and archistriatum. The latter will not be included in the present report. At least part of the neostriatum of birds has been homologized with the intermediolateral area of the alligator, and the bulk of the alligator dorsolateral area compared to the various hyper-striatal areas of birds. No homology with a structure in the brains of reptiles or mammals has been suggested for the ectostriatum (HUBER and CROSBY, 1929).

In histochemical preparations, the neostriatum and the main part of the hyper-striatum of the parakeet resembled the dorsolateral and intermediolateral areas of the alligator in showing weak to moderate SDH activity in the bulk of the neuropil (Fig. 33). A more strongly stained sheet of neuropil appeared in the

striatum, but insignificantly so in comparison with the homologous area in the turtle brain. Within the forebrain bundles may be seen the deeply stained cells of the globus pallidus (also see Figs. 22 and 23). Dorsal to the paleostriatum augmentatum lies the enormously enzymatically active area given the name "oval nucleus" in the present report. Medial to this, the dorsal striatum shows weak activity. The latter includes, in Fig. 16, the intermedio-lateral area (see Fig. 24). Figs. 17 and 18 show the "tail" of more deeply stained neuropil which comes to separate a large-celled area below from a smaller-celled area above. The latter dwindles caudally, as the former enlarges. In Fig. 18, weak enzymatic activity is seen in the neuropil of the main mass of the caudal dorso-lateral area. Diencephalic structures showing great enzymatic differentiation in Fig. 17 and 18 will be described in a future report, as will cortical structures such as the hippocampal formation. ×8

Fig. 19. Caiman. Sagittal section, SDH. A section near the midline of the brain to illustrate the relations of the "tail" to the mass of the dorsolateral area. Structures showing especially strong enzymatic activity in this section include the caudal hippocampal formation, habenular nucleus, nucleus reuniens (thalamus), cerebellar cortex and oculomotor and interpeduncular nuclei. ×8

Fig. 20. Caiman. Sagittal section, SDH. The intensely stained "oval nucleus" appears in rostral dorso-lateral area. Below it the moderately stained paleostriatum augmentatum caps the fanned-out portion of the lateral forebrain bundle. Among the fibers of the latter, the strongly active neurons of the globus pallidus are prominent, even at this low magnification. The nucleus of the lateral olfactory tract forms a prominent teardrop-shaped mass of moderately active neuropil wedged between the "oval nucleus" and the paleostriatum at the rostral pole of the telencephalon. Other prominently stained structures appearing in this section include optic tectum, inferior colliculus, cerebellum, ganglion isthmi, pretectal nucleus and nucleus of the basal optic root, forming a small ventral bulge. ×8

Fig. 21. Caiman. Sagittal section, AcPase. Section adjacent to that in Fig. 20. Attention is drawn especially to the stronger activity in neurons of the "oval nucleus", and the back-ground "browning" in this area. Such browning also appears in the optic tract, optic tectum, nucleus of the basal optic root, ganglion isthmi, and cerebellum. The area of the paleostriatum augmentatum can be seen to contain smaller and more weakly stained cells than the dorsal striatum. ×8

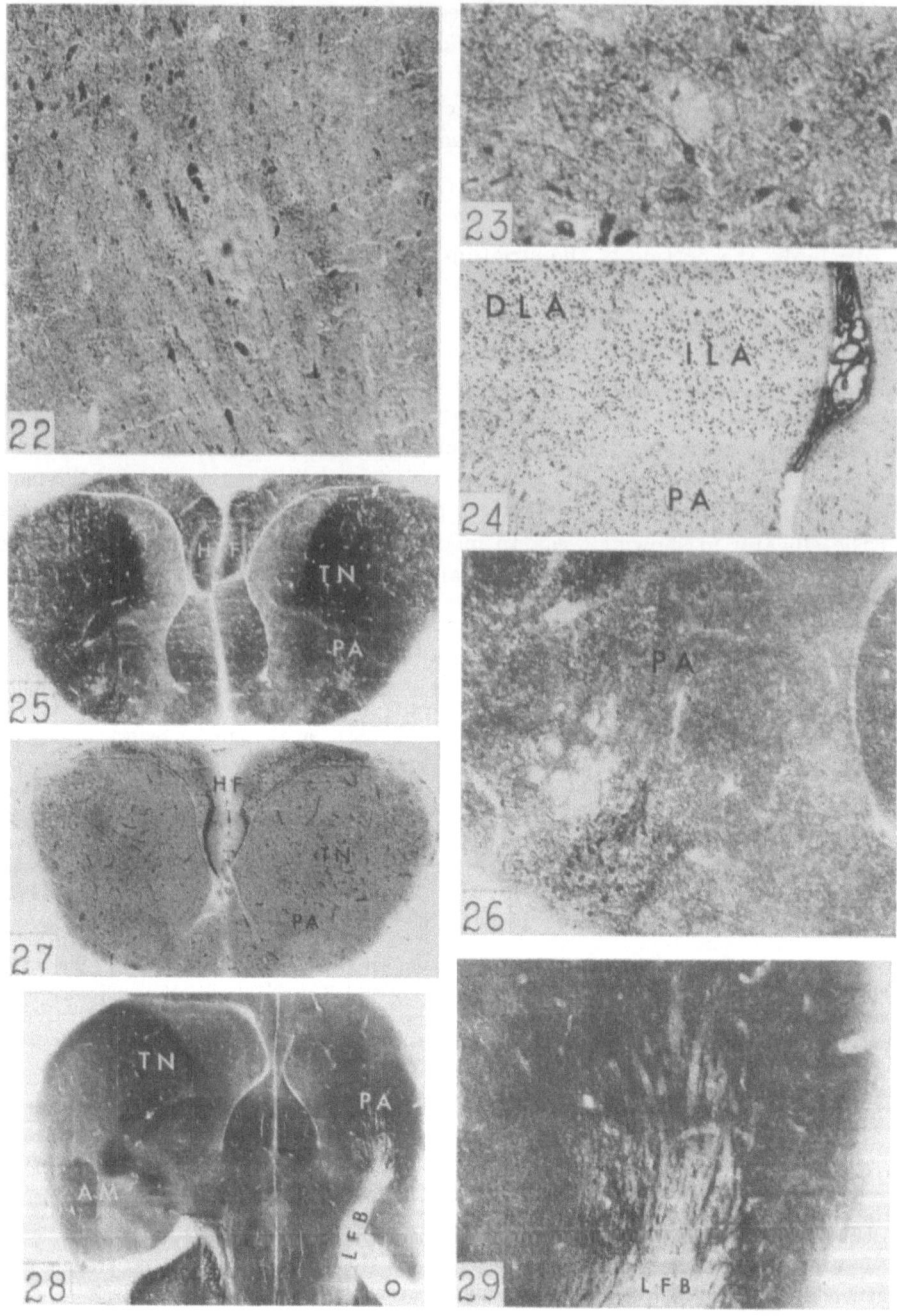

Fig. 22—29

Fig. 22. Caiman. Sagittal section, SDH. Strong activity is seen in neurons and processes of the globus pallidus lying among the fiber bundles of the lateral forebrain bundle. ×80

Fig. 23. Caiman. Horizontal section, SDH. A higher power view of part of the globus pallidus, showing strong enzymatic activity in the neuron bodies and processes. The light areas represent cross-sections through fiber bundles of the lateral forebrain bundle. ×180

medial neostriatum adjacent to the fibers of the lamina medullaris, which separates the palaeostriatum augmentatum from surrounding areas. In addition, a layer of neuropil intervening between the neostriatum and hyperstriatum showed stronger SDH activity. This was especially deeply stained and thicker close to the lateral ventricle (Figs. 33, 35). Both of these layers of neuropil showed "browning" in AcPase preparations. In sagittal sections both could be seen to extend into the caudal striatum above and behind the palaeostriatum augmentatum. Thus, both structures showed resemblances to the "tail" of more active neuropil which extended from the "oval nucleus" to the ventricle in SDH preparations of the caiman brain.

Neuron bodies in the parakeet neo- and hyperstriatal areas were not as prominently stained for SDH as those in the dorsolateral area of the alligator, or the DVR of the turtle. Like the palaeostriatum augmentatum, both the neo- and hyperstriatum of the parakeet contained a scattering of neurons with high AcPase activity, but most of the cells were not very active for this enzyme (Fig. 37). The frequency of neurons showing heavy AcPase staining appeared to be greater in the palaeostriatum augmentatum than in the dorsal areas. This was the reverse of the case with respect to the "putamen" and dorsal striatum in the reptiles.

Fig. 24. Caiman. Coronal section, AcPase. A rostral section through the telencephalon showing the especially high enzymatic activity in the large neurons of the intermediolateral area. Neurons of the dorsolateral area also are large and strongly stained, but those of the palaeostriatum augmentatum are small and weakly active. ×28.8

Fig. 25. *Anolis*. Coronal section, SDH. The most prominent enzymatically active part of the hemisphere is the "triangular nucleus". Below it lies the moderately strongly stained palaeostriatum augmentatum (or lateral palaeostriatum). Both show much greater activity than the medial part of the dorsal striatum. Within the palaeostriatum, light areas represent fiber bundles of the lateral forebrain bundle. A higher power view of the palaeostriatal area in this section appears in Fig. 26. ×20

Fig. 26. *Anolis*. Coronal section, SDH. A higher power view of part of the section shown in Fig. 25. Globus pallidus neurons, associated with the unstained fiber bundles of the lateral forebrain bundle, are intensely stained for SDH. Within the moderately stained neuropil of the palaeostriatum augmentatum, a scattering of strongly active neurons may be seen. The dark area above is the base of the "triangular nucleus". ×80

Fig. 27. *Anolis*. Coronal section, AcPase. A section sligthly rostral to that shown in Fig. 25. In contrast to the other species of animals studied, AcPase incubations of *Anolis* brain sections led to strong staining of blood vessels. This enables one to note that blood vessel density in the region of the "triangular nucleus" is greater than in other parts of the striatum (compare with Fig. 25). ×20

Fig. 28. *Anolis*. Horizontal section, SDH. As this section was cut obliquely, on the left side it shows the intensely stained "triangular nucleus", and on the right, the less strongly stained, ovoid-shaped palaeostriatum augmentatum, with the attached lateral forebrain bundle. A higher power view of the latter structures is shown in Fig. 29. The dark-stained structures in the caudal part of the left hemisphere are elements of the amygdala. ×20

Fig. 29. *Anolis*. Horizontal section, SDH. A higher power view of part of the section shown in Fig. 28. The deeply stained neurons of the globus pallidus are visible in association with fiber bundles of the lateral forebrain bundle. Within the rounded palaeostriatum augmentatum, a scattering of neurons with intense SDH activity may be seen embedded in the less strongly stained neuropil. ×80

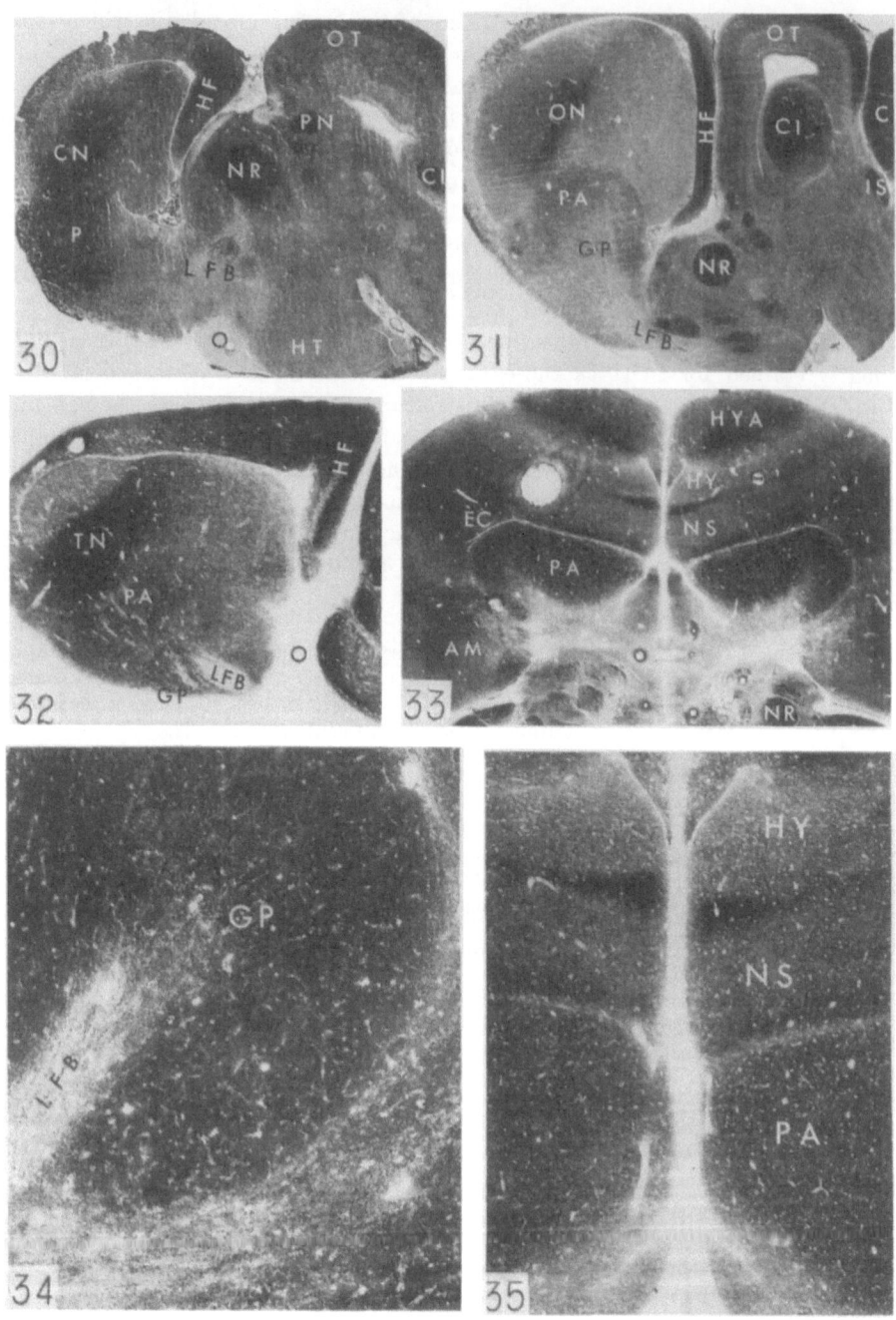

Fig. 30—35

Figs. 30—32 show comparable sections, at levels selected for comparison of the striatal structures described in this report, of turtle, caiman and lizard brains. Fig. 30. Western Painted Turtle. Sagittal section, SDH. ×11.5. Fig. 31. Caiman. Sagittal section, SDH. ×8. Fig. 32. *Anolis*. Sagittal section, SDH. ×25. In all three species, an enzymatically highly active area is found in the rostral dorsal striatum. In the turtle, this is the core nucleus of the dorsal ventricular ridge, in the alligator, it is the "oval nucleus", and in *Anolis*, it is the

(continued page 29)

The hyperstriatum accessorium of the parakeet contained very strong SDH activity (Fig. 33) and perhaps more neurons with strong AcPase staining than the main hyperstriatum or neostriatum. Because its homologies have been suggested to lie with certain cortical structures in reptiles and mammals (ARIENS KAPPERS et al., 1960; CROSBY et al., 1966), it will not be discussed further in the present report.

The ectostriatum of the parakeet lies rather higher in the hemisphere than it does in birds such as the sparrow (HUBER and CROSBY, 1929) or hummingbird (CRAIGIE, 1932). Because of this position and because of its appearance in histochemical preparations, it was strongly reminiscent of the "oval nucleus" in the caiman brain. Its central basal portion, adjoining the lamina medullaris and palaeostriatum augmentatum, was pale in SDH preparations, owing to the dense collection of fibers in that part (components of the lateral forebrain bundle — HUBER and CROSBY, 1929). But the main mass of this nucleus, as well as an area surrounding it, probably belonging to the neostriatum, contained highly intense SDH activity. This was present in both neuropil and neuron bodies (Figs. 33, 36). The fibrous core of this nucleus was sharply delineated in AcPase preparations by dense "browning", and many of the neurons showed intense AcPase activity (Fig. 37). Light NAD diaphorase staining in neuropil throughout the telencephalon, in sections thus far prepared, permitted clear visualization of the prevalence of highly active neurons in the ectostriatum, in contrast to their scarcity

"triangular nucleus". Adjacent to these ventrally is the palaeostriatum augmentum (or putamen of JOHNSTON, — in the turtle). This shows varying enzymatic activity in the three species, being most deeply stained in the turtle, and least so in the alligator. In Fig. 31 and 32, the lateral forebrain bundle is seen entering the palaeostriatum augmentum, and has associated with it the strongly active neurons of the globus pallidus. The turtle section does not show the latter features, as it was selected to show best the spatial relationship of the core nucleus to the rostral part of JOHNSTON's putamen. The globus pallidus and lateral forebrain bundle of the turtle in a more lateral sagittal section are seen in Fig. 3 to 5. Other structures showing especially high enzyme activity include hippocampal formation, nucleus rotundus, pretectal nucleus, optic tectum, inferior colliculus, cerebellum and nucleus isthmi

Fig. 33. Parakeet. Coronal section, SDH. The palaeostriatum augmentatum shows intense enzymatic activity in its neuropil. It is outlined by the unstained lamina medullaris, which separates it from the neostriatum medially and the ectostriatum laterally. The ectostriatum proper is surrounded by a "halo" of intensely stained neuropil, which actually occupies a part of the neostriatum. The neostriatum and hyperstriatum, both of which show rather weak enzymatic activity, are separated by a strip of strongly stained neuropil which becomes wider and more intensely stained near the lateral ventricle. The deeply stained dorsal area is the hyperstriatum accessorium. Below the ectostriatum, in the lateral part of the hemisphere, the dark-staining areas represent parts of the amygdala. The most prominent structures in the ventral region of the section are the optic lobes and the nucleus rotundus of the right side. $\times 8$

Fig. 34. Parakeet. Coronal section, SDH. Strongly stained neurons of the globus pallidus appear among the neuropil at the termination, as a distinct entity, of the lateral forebrain bundle. $\times 45$

Fig. 35. Parakeet. Coronal section, SDH. Strongest activity here is shown by the strip of neuropil separating the neostriatum from the hyperstriatum. Enzymatic activity is also intense in the palaeostriatum augmentatum and a strip of neostriatal neuropil adjacent to the unstained lamina medullaris. The fiber tracts running down from the hippocampal region are the septomesencephalic tracts. $\times 25$

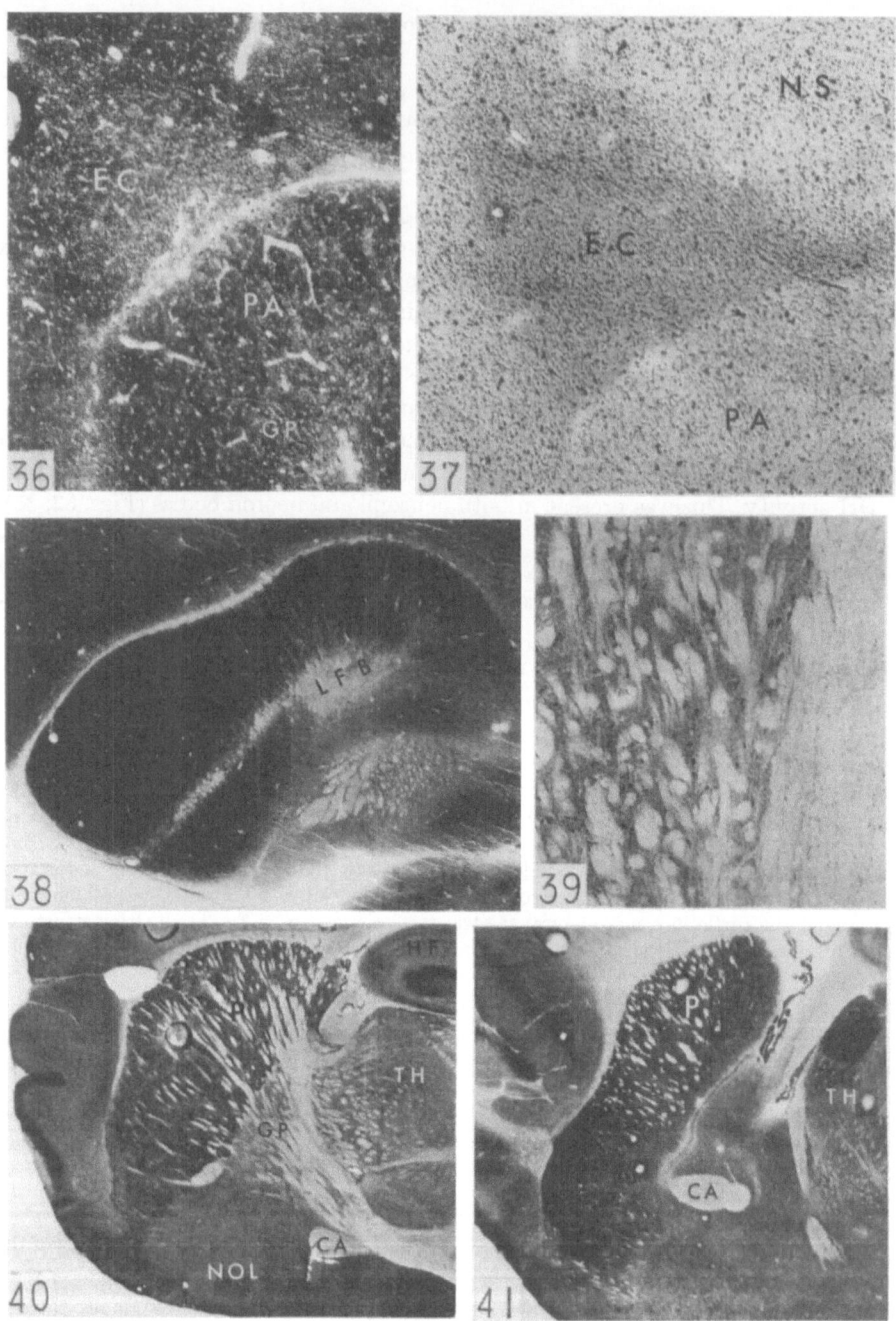

Fig. 36—41

Fig. 36. Parakeet. Coronal section, SDH. The palaeostriatum augmentatum shows strong
enzymatic activity in its neuropil. Enclosed within it, at the extreme lower right, a group
of the intensely stained neurons of the globus pallidus appears, with unstained fascicles of
the lateral forebrain bundle. Within the ectostriatum deeply stained neurons are visible,
and some strongly active cells are seen in the region of connection between the ectostriatum
and paleostriatum augmentatum. The fibrous core of the ectostriatum (see Fig. 37) is surrounded by an area of intensely stained neuropil which evidently lies in the neostriatum. ×45

or absence in all other striatal areas — with the exception of the globus pallidus and certain archistriatal parts. This differed somewhat from the SDH picture — for neurons with strong staining for SDH were found in the palaeostriatum augmentatum.

These observations on the histochemistry of the ectostriatum resembled those made, not only on the "oval nucleus" of the caiman brain, but also on the core nucleus of the turtle, the "triangular nucleus" of *Anolis*, and the "band nucleus" of *Phrynosoma*. The topographical relations of these various structures to the palaeostriatum augmentatum or putamen (of JOHNSTON), as seen in transverse and sagittal sections, also are similar (Figs. 1, 2). Species differences in the position, size and shape of these reptilian structures is equalled by variations in the position, size and shape of the ectostriatum in different species of birds (HUBER and CROSBY, 1929; CRAIGIE, 1932; DURWARD, 1932; REVZIN and KARTEN, 1967). It seems possible that all of these structures in the reptile and bird brains may be homologues.

Discussion

A. Histochemical Factors

Before discussion of the comparative neurological aspects of the chemo-architectonic observations presented here, it is important to mention a few points with respect to the more general interpretation, and limitations, of these observations.

Both biochemical and histochemical studies on the brain have shown that succinic dehydrogenase is largely confined to mitochondria (ABOOD et al., 1952; FRIEDE and PAX, 1961). The latter are present in dendritic and synaptic structures (neuropil) and in neuron perikarya. However at the level of resolution employed in the work reported here, SDH activity in neuron perikaryal cytoplasm is indistinguishable from that which might be present in synaptic boutons

Fig. 37. Parakeet. Coronal section, AcPase. The fibrous core of the ectostriatum shows prominent "browning" in this lead sulfide preparation. Larger and strongly stained neurons are present among the generally smaller celled populations in the neostriatum, palaeostriatum augmentatum and archistriatum (lower left), as well as in the ectostriatum itself. A certain degree of cellular continuity appears to be present between the ectostriatum and palaeostriatum augmentatum, where fiber tracts pass across the boundary. ×45

Fig. 38. Parakeet. Sagittal section, SDH. Intense enzymatic activity is shown by the palaeostriatum augmentatum. The unstained fiber bundles within this structure are elements of the lateral forebrain bundle, and those appearing below it are related to the archistriatum (amygdalae) and the anterior commissure. ×16

Fig. 39. Mouse. Sagittal section, SDH. The globus pallidus shows strongly stained neurons and rather weakly active neuropil. ×85

Fig. 40. Mouse. Sagittal section, NAD diaphorase. The putamen shows stronger activity than most other brain structures. Exceptions which show here are the tuberculum olfactorium (extreme left) and parts of hippocampus. ×16

Fig. 41. Mouse. Sagittal section, SDH. The putamen-caudate nucleus (P), merging with the nucleus accumbens rostro-ventrally, is more intensely stained for this enzyme than it is for NAD diaphorase. The anterior commissure is unstained. The deeply stained nucleus in the thalamus at the right is the nucleus anterior dorsalis. ×18

at the surface of the cell body. Therefore statements regarding the strength of the somatic reaction for this enzyme should be viewed with this fact in mind. The same qualification applies to NAD diaphorase activity, as it also is a mitochondrial enzyme.

Acid phosphatase in neurons is present in lysosomes and the Golgi apparatus. The latter is a localized organelle with a more restricted AcPase distribution than in lysosomes (GOLDFISCHER et al., 1964). Lysosomes appear to be randomly distributed in the cytoplasm of neurons, and are commonly seen in the dendrites. They are generally more numerous in larger neurons (BECKER et al., 1960). Since, in the present acid phosphatase preparations, cell boundaries, especially of smaller neurons, were often not readily distinguished, estimations of cell size could be influenced by intracellular aggregation of stained components. Therefore descriptions of relative cell sizes based on study of these preparations may be very rough approximations in comparison with cytological descriptions based on classical neurohistological material.

The attempt has been made here to express comparisons of enzymatic activities of structures in the brains of different species in *relative* terms. Thus, the enzyme activity in a particular structure was compared in two specimens, by bearing in mind its staining intensity in relation to the overall picture in each specimen. Several specimens of each animal type were prepared to demonstrate any one enzyme, so that observations were never based on an isolated and possibly aberrant preparation. True direct comparisons between species would necessitate community incubations of contemporaneously prepared specimens. Such preparations permitting direct comparisons will be included in future studies. However, comparisons so based may be subject to at least two further sources of error in comparative interpretations: 1. Since the histochemical methods applied were developed for use on mammalian material, the possibility exists that the conditions of incubation (pH, temperature, etc.) might not be optimal for submammalian material; 2. It is known that in mammals the intensity of overall enzymatic activity, especially of oxidative enzymes, shows an inverse relationship to brain size (FRIEDE, 1965). A possible parallel relationship within reptilia may be noted in some of the illustrations accompanying the present report. Within the same period of incubation, sections of the small *Anolis* brain became more heavily stained for SDH than did those of the much larger young caiman brain. FRIEDE has pointed out that this factor constitutes a strong argument in favor of the approach of making comparisons based on relative *patterns* of activity, rather than on absolute values for a given structure.

Apart from the functional significance of relative enzymic activities in structures of the submammalian brain, it may be noted that some of the histochemical procedures applied in this investigation delineated certain structures much more strikingly than have standard neurohistological methods. For this reason, the use of such histochemical methods could be helpful in comparative neurology, regardless of the particular interest in biochemical factors.

B. Chemoarchitectonics and Comparative Neurology

It is generally agreed that the palaeostriatum of submammals includes the globus pallidus and the part of the head of the caudate nucleus associated with the nucleus accumbens in mammals (JOHNSTON, 1923; CROSBY et al., 1966). However most students of the reptilian or avian brain have not followed JOHNSTON in equating a part of the so-called palaeostriatum with the putamen of mammals. Rather, the latter generally has been categorized as a neostriatal structure. Although ELLIOT SMITH (1919) concurred with JOHNSTON in the assignment of the intraventricular bulge of the DVR to part of the caudate nucleus of mammals, he located his homologue of the putamen in the reptile brain in a position dorsal to that of JOHNSTON's palaeostrial putamen. This localization, when applied to the turtle brain, necessarily would have had to coincide with at least part of the core nucleus of the DVR, whereas JOHNSTON believed the DVR as a whole to represent the caudate nucleus. In a recent paper, CROSBY et al.,(1966) homologized the palaeostriatum augmentatum and neostriatum of submammals with parts of the caudate nucleus, but the mammalian putamen was discussed only as possibly having a partial representation in the neostriatum, and perhaps some relationship with the hyperstriatum. In a photograph of a turtle brain section, rostral to the core nucleus, the area of JOHNSTON's putamen was labelled "neostriatum".

HOLMGREN (1925) is one of the few students of the reptilian brain who agreed with JOHNSTON in assigning the putamen to the palaeostriatum (or "striatum proper"). However, he differed from JOHNSTON in concluding that the caudate nucleus was represented not in the dorsal ventricular ridge, but in the palaeostriatum, together with the putamen. The only mammalian homology that he accepted for the DVR was with the claustrum. Also, he believed the globus pallidus to be represented in JOHNSTON's stria bed, and not in the large cells associated with the lateral forebrain bundle.

These few examples, together with those previously mentioned in introducing the observations in this report, perhaps will suffice to illustrate not only the variety of interpretations developed by various authors with respect to the representation of the parts of the mammalian basal ganglia in submammalian brains, but the differences in terminology, having different connotations, applied to the same structures. Aside from the question of mammalian homology, some new means of equating areas between one submammalian type and another seems to be called for.

The histochemical observations described in the present report appear to show a consistant pattern for several striatal components in four species of reptiles and a bird. Certain of these components have not been described before. As a prelude to further discussion, a short review of this pattern may be presented in four main divisions:

1. All animals possessed a group of large neurons associated with the telencephalic portion of the lateral forebrain bundle. These were identifiable as a globus pallidus. These neurons in all species showed similar oxidative enzymatic characteristics to those of the globus pallidus of mammals, although they were more variable in their acid phosphatase activity.

2. In every species the lateral forebrain bundle spread into a basal forebrain area with similar enzymatic and other characteristics. These included: a population of preponderently small cells with weak acid phosphatase activity; a minor component of larger cells with strong oxidative enzymatic activity; and a relatively strong oxidative enzymatic staining of the neuropil. The latter was least

marked in the caiman and greatest in the turtle and parakeet. In spite of the latter variation, these structures in the various forms were so similar in these features, as well as in topographical relations, that there seems little doubt but that they are homologous. The cytological and histochemical characteristics of these structures, especially as seen in the turtle, were similar to those of the putamen-caudate nucleus complex of mammals, as described in the literature, and as observed here in the mouse. In the turtle brain, this area is the putamen of Johnston; in the lizards it is the lateral palaeostriatum, or somatic striatum, of various authors; in the caiman brain it is included in Crosby's ventrolateral small-celled area; and in the parakeet, it is the palaeostriatum augmentatum.

3. All forms possessed a more or less well-defined area in the rostral dorso-lateral striatum which contained strong oxidative enzymatic activity in its neuro-pil. All of these areas also contained neurons with both strong somatic oxidative enzyme staining and high acid phosphatase activity. In all, a diffuse brown staining (probably of small fibers) was prominent in preparations involving lead sulfide deposition. In the turtle, this area was Johnston's core nucleus of the dorsal ventricular ridge, and in the parakeet, it was the ectostriatum. Both of these are cytologically easily distinguishable, but the areas of this type in the lizard and caiman brains would not be differentiated in ordinary histological preparations, and have not been described in the literature. Although there was a good deal of variation in the size and shape of these various structures, all had a similar topographical relation to the underlying palaeostriatum, and in the turtle and caiman, a similar position with respect to the nucleus of the lateral olfactory tract and piriform cortex. In the turtle, the core nucleus could be seen to lie in close relation to the "pallial thickening" of Johnston. It was suggested that these structures in the four reptiles were homologous with one another, and per-haps also with the ectostriatum of birds.

The relative oxidative enzymatic activity shown by these dorsal areas in each species appeared to be inversely related to the activity shown by the palaeo-striatum. Thus, in the caiman brain, in which the palaeostriatum showed com-paratively the least activity, the dorsal "oval area" was comparatively the most intensely stained for SDH. The reverse was true of the turtle brain.

4. The rostral striatal areas medial to, and/or ventromedial to, the structures just described, contained in all forms relatively weak oxidative enzymatic activity in the neuropil. Within this sector of the striatum, especially strong acid phos-phatase staining occurred in neurons of the intermediolateral area of the caiman, and of probably corresponding areas in the lizard brains. All neurons of the antero-medial part of the DVR in the turtle contained very strong acid phosphatase activity. In the parakeet, both the neostriatum and hyperstriatum medial to the ectostriatum contained a mixed population of cells with respect to acid phos-phatase activity; the majority were not strongly stained.

In sum: Palaeostriatal structures in submammals were found to show close histochemical and cytological similarities to the putamen-caudate nucleus complex and globus pallidus of mammals. Much less resemblance was found between any part of the dorsal striatum in the submammals and the putamen-caudate complex of mammals. These observations could be interpreted in favor of Johnston's view that the palaeostriatum includes the putamen as well as the head of the caudate nucleus, especially when topographical relations to the globus pallidus are considered. If this were the case, the main part (tail) of the caudate then would remain the sole mammalian "neostriatal" structure whose submammalian representation appeared in the dorsal area.

It was suggested above that the histochemical activity of the core nucleus of the DVR in the turtle might indicate that it made a better candidate for a homologue of the caudate nucleus of mammals than would the DVR taken as a whole, as proposed by JOHNSTON (1915; 1923). The presence of metabolically similar dorsolaterally located areas in the brains of lizards and caiman, in combination with published observations on cortico-striatal relationships in reptile brains, made it seem not illogical that these areas could correspond in a functional way to the caudate nucleus of mammals, even if other parts might have to be included in a strictly structural evolutionary homology. CROSBY (1917) concluded that most of the anterior dorsolateral area in the alligator had exclusively somatic sensory (thalamic) connections, and was closely related to the anterior infolding of the dorso-lateral cortex which she called "primordial general cortex", and which JOHNSTON called the "pallial thickening". In *Chameleon*, SHANKLIN'S (1930) "neostriatum pars lateralis" occupied an area whose medial boundary appears to correspond to that of the "triangular nucleus" in *Anolis*. SHANKLIN found this lateral part of the neostriatum to be thalamically connected, whereas he was unable to follow lateral forebrain bundle fibers into his "neostriatum pars medialis". Like CROSBY, he found a close dorsolateral contact of his pars lateralis with the infolded "primordium of the general cortex", especially rostrally. The same close relationship is found between the pallial thickening and core nucleus in the turtle. The following excerpt from CROSBY'S (1917) paper bears on this point:

"... at the level of the primordial general cortex, this dorsal area is entered almost exclusively by the somatic correlation fibers and hence is a somatic correlation center of striatal type. This area at its anterior end probably exhibits the highest type of somatic correlation tissue found in the brain of the alligator, and the entrance of association fibers from the adjacent cortical centers into its dorsal part has given the conditions favorable for the differentiation of primordial general cortex (i.e. cortex approaching the neopallial in type)."

Support for a more widespread representation of the caudate nucleus in the submammalian dorsal striatum, rather than its restriction to parts showing especially high oxidative enzyme activities, may be found in FRIEDE'S (1965) observation of extensive fluctuation of enzymatic activity levels in various portions of the caudate nucleus of mammals. Of probably more weighty significance are the findings of BERTLER *et al.* (1964) in the pigeon brain. These investigators, in studying the localization of monoamines by fluorescence histochemical methods, discovered a concentration of closely packed dopamine-containing fibers in both the palaeostriatum augmentatum and "at least the main part of the neostriatum" (the ectostriatum was not mentioned). The detailed cytological observations on this system in pigeon brain were similar to those made on the putamen-caudate nucleus complex in mammalian brain.

Possible homology between the dorsolateral enzymatically specialized striatal areas of reptiles (core nucleus, etc.) and the ectostriatum of birds was suggested above. Implications of this idea must be considered in relation to the idea of caudate nucleus localization just discussed in connection with these reptilian areas. All of these structures in reptile and bird brains, whether or not specifically identified in the anatomical literature, share the telencephalic distribution of the lateral forebrain bundle (Crosby, 1917; Huber and Crosby, 1929; Shanklin, 1930). Recently, Revzin and Karten (1967), using electrophysiological methods in the pigeon brain, have demonstrated a major (and perhaps exclusive) projection of the nucleus rotundus of the thalamus to the ectostriatum. In the light of their earlier conclusions with respect to the tecto-rotundal pathway (Karten and Revzin, 1966), these authors believed that the tecto-rotundo-ectostriatal pathway in birds subserves vision, and might be the major route of transmission of visual information to the telencephalon. However, these conclusions may be open to criticism related to the interpretation which these investigators placed on some of their observations. In their 1966 publication, based on the application of Nauta-Gygax methods to the pigeon brain, they reported that large numbers of degenerating fibers were found in the nucleus rotundus after tectal lesions, but degeneration did not occur in that nucleus after lesions of spinal and medullary sensory centers or deep cerebellar nuclei. They interpreted these observations as evidence that the nucleus rotundus could not be homologous with the ventral nuclear complex of the mammalian thalamus, which is the primary recipient of such lower sensory and cerebellar impulses. However, it has been repeatedly emphasized by Huber and Crosby (1926, 1933, 1943) that a characteristic pattern in the reptilian brain is the convergence of somatic sensory impulses from spinal, medullary and brainstem centers upon the highly differentiated correlation center developed in the optic tectum. After synapse, many of the impulses are projected forward upon the dorsal thalamus; the nucleus rotundus is one of the chief recipients of the large tecto-thalamic tracts. Tecto-thalamic and thalamo-tectal connections in birds are similar to those of alligator (Huber and Crosby, 1929). If these statements are accepted, then without the occurrence of trans-synaptic degeneration, it would not be expected that lesions of lower sensory centers in reptiles, and presumably birds, would result in appreciable axon degeneration in the dorsal thalamus itself. Hence, Karten and Revzin's basis for exclusion of the nucleus rotundus as a ventrobasal homologue may not be valid. This in turn could re-open the possibility that the projection of the nucleus rotundus to the ectostriatum might include all somatic sensory modalities, rather than exclusively visual ones. If this were so, the ectostriatum could be homologous with some part of the mammalian caudate nucleus-putamen complex, rather than perhaps a specialized structure peculiar to the reptilian-avian line. In mammals, although cortical efferents are predominant, the ventral thalamic nucleus also retains efferent connections with the putamen and caudate nucleus (Crosby, Humphrey and Lauer, 1962)

In both the caiman and the lizards, large-celled areas in the medial anterior dorsolateral area showed histochemical, as well as topographical similarities to the antero-medial part of the DVR in the turtle. Like that part of the turtle brain, Crosby's intermediolateral area in the caiman contained large cells which stained strongly for AcPase and SDH, and a neuropil which was only moderately active for SDH. In the turtle, by cytological and histochemical criteria, the antero-medial DVR formed a continuum with the caudal part of the DVR. Crosby noted that her intermediolateral area appeared to be continuous with the caudal dorsolateral area, and probably was a part of it. In the two lizards, large-celled areas in the neostriatum, medial to the "triangular" or "band nucleus" areas of high SDH activity, and just dorsal to the palaeostriatum, showed histochemical characteristics very similar to those of the intermediolateral area of the caiman. These nuclei in the lizards corresponded to part of SHANKLIN's neostriatum pars medialis, which he stated to be continuous with his archistriatum pars medialis caudally. On the basis of the generally similar histochemical characteristics, and the anatomical analyses presented by CROSBY, SHANKLIN, and others, it seems likely that the antero-medial part of the DVR in the turtle and the intermediolateral areas of the caiman and lizards are comparable.

HUBER and CROSBY (1929) believed the intermediolateral area of the alligator to correspond with part at least of the anterior neostriatal area of birds. In the parakeet brain, the generally rather weak staining for SDH shown by the neuropil of the neostriatum resembled that of the intermediolateral areas of caiman and lizards. In dehydrogenase preparations, both the caiman and parakeet brains showed a narrow, more deeply stained layer of neuropil separating dorsal and ventral parts of the medial dorsal striatum. This structure ("tail") in the caiman separated a large-celled area below, which was continuous with the intermediolateral area, from a smaller-celled area above. In the parakeet brain, a similar-appearing structure separated the neostriatum below, from the hyperstriatum above. If the intermediolateral area of crocodilians represents part of the neostriatum of birds, the "tail" in the caiman may correspond in functional significance with the layer described in the parakeet brain. Reciprocally, the occurrence of these similar enzymatically active layers in similar locations in brains of caiman and parakeet lends support to HUBER and CROSBY's analysis of homology.

Exclusive of the intermediolateral area, the rostral neostriatum medial to the "oval nucleus" in the caiman, and corresponding parts in lizards, showed poor SDH in the neuropil, and smaller cell bodies. These showed relatively weak AcPase, as well as SDH, staining. A histochemically corresponding area was not discovered in the turtle brain. In the caiman, this area consisted of the region above the "tail" which appeared in dehydrogenase preparations. This dorsomedial histochemically distinguishable region in the caiman brain could correspond to a part believed by CROSBY (1917) to be the only exclusively somatically connected portion of the dorsolateral area which extends into the caudal region. She did not, however, show a cytologically distinguishable corresponding area in her figures. If, as discussed in the previous paragraph, the "tail" structures observed in caiman and parakeet brains have similar significance, the part of the brain dorsal to the "tail" in the caiman would correspond with part of the hyperstriatum of birds.

How do these histochemical findings on submammalian brains fare in response to the two questions presented in the introduction to this report? A positive answer to the first question in part at least is indicated by observations made on the globus pallidus and the palaeostriatum augmentatum. Regardless of the question of its incorporation of the homologue of the putamen, the histochemical characteristics of the palaeostriatum augmentatum are, in general, similar to those of the head of the caudate nucleus of mammals. This is true, not only with respect to the enzyme activities reported here, but also to catechol amine distribution, as cited above. However, it is clear that although histochemical activities of the general metabolic type as described here may be suggestive of homologies, in their implications of functional similarity, more physiologically specific methods may be required to either solidify or negate such suggestions. This is certainly true of the main mass of the dorsal striatum in reptiles and birds, in which definitive histochemical resemblances to parts of the basal ganglia of mammals seem so far to be largely confined to catechol amine activity. (That such a general limitation may not be required for many parts of the brain will become more evident in forthcoming parts of this series.)

The divergent evolutionary development of higher correlative centers, structurally represented by the enlargement of the dorsal striatum in the reptilian-avian line, and by the enlargement and differentiation of the cerebral cortex in the mammalian line (HUBER and CROSBY, 1929) might naturally be expected to lead to widely differing overall biochemical patterns in the "residual" striatum of mammals and that with a great new superstructure built into it, as found in birds. In keeping with this, the dorsal striatum of the most bird-like reptile (caiman) showed more, and larger, histochemically differentiated areas having little resemblance to the mammalian putamen-caudate complex, than did the dorsal striatum of the most primitive reptile (turtle). As a corollary of this, JOHNSTON's putamen in the turtle brain, which showed strong histochemical resemblances to the mammalian putamen-caudate complex, is proportionately larger than the corresponding palaeostriatal area in the caiman and parakeet.

Summary

Chemoarchitectonic observations were made on the striatum in turtle, Carolina Anole, Horned Lizard, young caiman and parakeet. Enzyme histochemical observations on these forms were compared with those on the basal ganglia of mammals. Enzyme activities studied included succinate dehydrogenase, reduced nicotinamide-adenine dinucleotide tetrazolium reductase, acid phosphatase and thiamine pyrophosphatase.

The globus pallidus in the submammalian forms was similar to that of mammals with respect to the activities of the enzymes studied.

Ventral striatal areas, previously termed "putamen" (turtle), "somatic striatum" (lizards), "ventro-lateral small-celled area" (alligator) and "paleostriatum augmentatum" (birds), showed similar enzymatic characteristics, which resembled those of the putamen-caudate complex of mammals. The latter correspondence was greatest in turtles. In the lizards and caiman, dehydrogenase activity in the neuropil of these structures was less than in the neuropil of the mammalian structures. Similarities in enzymatic characteritics suggest homology between these structures in reptiles and bird.

Dehydrogenase activity was intense in an area of the reptilian dorsolateral striatum. The large neurons in this area showed high acid phosphatase. This area in turtles constitutes the cytologically distinct "core nucleus of the dorsal ventricular ridge". In lizards and caiman, these areas were not cytologically distinct. The ectostriatum of parakeet showed histochemical similarities to these areas in the reptiles.

The intermediolateral area of caiman, medial portions of lizard striatum, and the rostro-medial part of the dorsal ventricular ridge in turtle had similar enzymatic activities. These included relatively weak dehydrogenase activity in the neuropil and particularly strong acid phosphatase activity in the large neurons.

These areas in reptiles also showed enzymatic resemblances to parts of the neo-striatum of parakeet.

A dorsomedial sector of the striatum in caiman and lizards showed generally weak enzyme activities and contained relatively small cells. The similar appearance of certain histochemically demonstrable landmarks in caiman and parakeet striata suggested homology of the aforementioned areas with the avian hyper-striatum. No comparable area was identified in the turtle.

Acknowledgements. I wish to express my deep appreciation for the inspiration and encouragement given me by the late Dr. ERNST SCHARRER, who sponsored me in this project. To him belongs the credit for interesting me in the "chemoarchitectonics" of the brain, an area in which he had long had a special interest, and for which he himself had coined the name. I should like to dedicate to his memory a series of reports, of which this is the first, in which it is planned to summarize the histochemical comparisons which I have been able to make between reptilian and mammalian brains.

I also wish to convey my gratitude for the encouragement and instruction of the late Dr. HELEN DEANE. It was she who introduced me to the field of histochemistry, but also gave me all possible help in the development of the material and the communication of it.

I am much indebted to Dr. BERTA SCHARRER for making available to me the services of her technician, Mrs. CELIA GRUBMAN, and her laboratory facilities. I also wish to thank Mrs. GRUBMAN herself, for the preparation of most of the histochemical material, and for her patience in the face of numerous technical difficulties with the material. Appreciation is due to Misses MARGRIT KNITTER and MARIANNE VAN HOOREN for their help with the photography, and for the darkroom work on the illustrations.

Bibliography

ABOOD, L. G., R. W. GERARD, J. BANKS, and R. D. TSCHIRGI: Substrate and enzyme distribution in cells and cell fractions of nervous system. Amer. J. Physiol. **168**, 728—738 (1952).

ADAMS, C. W. M (ed.): Neurohistochemistry. Amsterdam: Elsevier 1965.

ARIENS KAPPERS, C. U., C. G. HUBER, and E. C. CROSBY: The comparative anatomy of the nervous system of the vertebrates, including man (reprint edition). New York: Hafner 1960.

ARMSTRONG, J. A., H. J. GAMBLE, and F. GOLBY: Observations on the olfactory apparatus and telencephalon of *Anolis*, a microsmatic lizard. J. Anat. (Lond.) **87**, 288—307 (1953).

BARKA, T., and P. J. ANDERSON: Histochemistry. New York: Harper & Row 1965.

BECKER, N. H., A. B. NOVIKOFF, and S. GOLDFISCHER: A cytochemical study of the neuronal Golgi apparatus. Arch. Neurol. (Chic.) **5**, 497—503 (1961).

BERTLER, Å., B. FALCK, C. G. GOTTFRIES, L. LJUNGGREN, and E. ROSENGREN: Some observations on adrenergic connections between mesencephalon and cerebral hemispheres. Acta pharmacol. (Kbh.) **21**, 283—289 (1964).

CAIRNEY, J.: A general survey of the forebrain of *Sphenodon punctatum*. J. comp. Neurol. **42**, 255—348 (1926).

CRAIGIE, E. H.: The cell structure of the cerebral hemisphere of the hummingbird. J. comp. Neurol. **56**, 135—168 (1932).

— Vascularization on the brains of reptiles. I. The painted turtle, *Chrysemis picta marginata* Agassiz. J. comp. Neurol. **74**, 247—264 (1941).

CROSBY, E. C.: The forebrain of *Alligator Mississippiensis*. J. comp. Neurol. **27**, 325—40? (1917).

— B. R. DEJONGE, and R. C. SCHNEIDER: Evidence for some of the trends in the phylogenetic development of the vertebrate telencephalon. In: Evolution of the forebrain (R. HASSLER and H. STEPHAN, eds.), p. 117—135. Stuttgart: Thieme 1966.

— T. HUMPHREY, and E. W. LAUER: Correlative anatomy of the nervous system. New York: The MacMillan Company 1962.

DEANE, H. W., B. L. RUBIN, E. C. DRIKS, B. L. LOBEL, and G. LEIPSNER: Trophoblastic giant cells in placentas of rats and mice and their probable role in steroidhormone production. Endocrinology **70**, 407—419 (1962).

DURWARD, A.: The cell masses in the forebrain of *Sphenodon punctatum*. J. Anat. (Lond.) **65** 8—44 (1930).

— Observations on the cell masses in the cerebral hemispheres of the New Zealand kiwi (*Apteryx australis*). J. Anat. (Lond.) **66**, 437—477 (1932).

ELLIOT SMITH, G.: A preliminary note on the morphology of the corpus striatum and the origin of the neopallium. J. Anat. (Lond.) **53**, 271—291 (1919).

FLEISCHHAUER, K., u. E. HORSTMANN: Intravitale Dithizonfärbung homologer Felder der Ammonsformation von Säugern. Z. Zellforsch. **46**, 598—609 (1957).

FOX, C. A., D. E. HILLMAN, K. A. SIEGESMUND, and L. A. SETHER: The primate globus pallidus and its feline and avian homologues: a Golgi and electron microscopic study. In: Evolution of the forebrain (R. HASSLER and H. STEPHEN, eds.), p. 237—248. Stuttgart: Thieme 1966.

FRIEDE, R. L.: Histochemical investigations on succinic dehydrogenase in the central nervous system. II. Atlas of the medulla oblongata of the guinea pig. J. Neurochem. **4**, 111—123 (1959a).

— Histochemical investigations on succinic dehydrogenase in the central nervous system. III. Atlas of the midbrain of the guinea pig, including pons and cerebellum. J. Neurochem. **4**, 290—303 (1959b).

— Histochemical investigations on succinic dehydrogenase in the central nervous system. V. The diencephalon and basal telencephalic centres of the guinea pig. J. Neurochem. **6**, 190—199 (1961a).

— A histochemical atlas of tissue oxidation in the brain stem of the cat. New York: Hafner 1961b.

— Topographic brain chemistry. New York: Academic Press 1966.

—, and L. M. FLEMING: A mapping of oxidative enzymes in the human brain. J. Neurochem. **9**, 179—198 (1962).

— — A mapping of the distribution of lactic dehydrogenase in the brain of the rhesus monkey. Amer. J. Anat. **113**, 215—234 (1963).

—, and M. KNOLLER: A quantitative mapping of acid phosphatase in the brain of the rhesus monkey. J. Neurochem. **12**, 441—450 (1965).

—, and R. A. PAX: Mitochondria and mitochondrial enzymes. A comparative study of localization in the cat's brain stem. Histochemie **2**, 186—191 (1961).

FUXE, K., and L. LJUNGGREN: Cellular localization of monoamines in the upper brain stem of the pigeon. J. comp. Neurol. **125**, 355—381 (1965).

GOLDBY, F.: The cerebral hemispheres of *Lacerta viridis*. J. Anat. (Lond.) **68**, 157—215 (1934).

GOLDFISCHER, S., E. ESSNER, and A. B. NOVIKOFF: The localization of phosphatase activities at the level of ultrastructure. J. Histochem. Cytochem. **12**, 72—95 (1964).

GOMORI, G.: Microscopic histochemistry. Chicago: University of Chicago Press 1952.

GUSEL'NIKOV, V. I., and A. Y. SUPIN: Visual and auditory regions in hemispheres of lizard forebrain. Fed. Proc. **23**, Translation Suppl.: T641—646 (1964).

HERIC, T. M., and L. KRUGER: The electrical response evoked in the reptilian optic tectum by afferent stimulation. Brain Res. **1**, 187—199 (1966).

HOLMGREN, N.: Points of view concerning forebrain morphology in higher vertebrates. Acta zool. (Stockh.) **6**, 413—477 (1925).

HUBER, G. C., and E. C. CROSBY: On thalamic and tectal nuclei and fiber paths in the brain of the American alligator. J. comp. Neurol. **40**, 97—227 (1926).

— — The nuclei and fiber paths of the avian diencephalon, with consideration of telencephalic and certain mesencephalic centers and connections. J. comp. Neurol. **48**, 1—225 (1929).

— — The reptilian optic tectum. J. comp. Neurol. **57**, 57—163 (1933).

— — A comparison of the mammalian and reptilian tecta. J. comp. Neurol. **78**, 133—168 (1943).

JOHNSTON, J. B.: The cell masses in the forebrain of the turtle, *Cistudo carolina*. J. comp. Neurol. **25**, 393—468 (1915).

— The development of the dorsal ventricular ridge in turtles. J. comp. Neurol. **26**, 481—505 (1916).

JOHNSTON, J. B.: Further contributions to the study of the evolution of the forebrain. J. comp. Neurol. **35**, 337—481 (1923).

KARAMYAN, A. I., and M. G. BELEKHOVA: Functional evolution of nonspecific thalamo-cortical system. Fed. Proc. **23**, Translation Suppl., Part II: T1189—1194 (1964).

KARTEN, H. J., and A. M. REVZIN: The afferent connections of the nucleus rotundus in the pigeon. Brain Res. **2**, 368—377 (1966).

KONIG, J. F. R., and R. A. KLIPPEL: The rat brain. A stereotaxic atlas. Baltimore: Williams & Wilkins Company 1963.

KRUGER, L., and E. C. BERKOWITZ: The main afferent connections of the reptilian telencephalon as determined by degeneration and electrophysiological methods. J. comp. Neurol. **115**, 125—142 (1960).

MASAI, H.: Comparative neurobiological studies on the glycogen distribution in the central nervous system of submammals. Yokohama med. Bull. **12**, 239—260 (1961 a).

— Glycogen distribution in the subfornical organ of submammals. Yokohama med. Bull. **12**, 261—264 (1961 b).

—, and S. MATANO: Comparative neurological studies on respiratory enzymic activity in the central nervous system of submammals. I. Birds. Yokohama med. Bull. **12**, 265—270 (1961 a).

— — Comparative neurological studies on respiratory enzymic activity in the central nervous system of submammals. II. Fishes and amphibia. Yokohama med. Bull. **12**, 271—276 (1961 b).

— T. KUSUNOKI, and H. ISHIBASHI: Comparative studies on the distribution of monoamine oxidase and succinic dehydrogenase in vertebrates' forebrain. In: Evolution of the forebrain (R. HASSLER and H. STEPHEN, eds.), p. 271—275. Stuttgart: Thieme 1966.

MEESSEN, H., and J. OLSZEWSKI: A cytoarchitectonic atlas of the rhombencephalon of the rabbit. Basel: Karger 1949.

NOVIKOFF, A. B., and S. GOLDFISCHER: Nucleosidediphosphatase activity in the Golgi apparatus and its usefulness for cytological studies. Proc. nat. Acad. Sci. (Wash.) **47**, 802—810 (1961).

ORTMANN, R.: Die Chemoarchitektonik des Gehirns. Dtsch. med. Wschr. **86**, 2063—2068 (1961).

PAPEZ, J. W. Thalamus of turtles and thalamic evolution. J. comp. Neurol. **61**, 433—475 (1935).

PEARSE, A. G. E.: Histochemistry, 2nd ed., p. 882. Boston: Little, Brown & Co. 1960.

POWELL, T. P. S., and W. M. COWAN: The thalamic projection upon the telencephalon in the pigeon (*Columba livia*). J. Anat. (Lond.) **95**, 78—109 (1961).

—, and L. KRUGER: The thalamic projection upon the telencephalon in *Lacerta viridis*. J. Anat. (Lond.) **94**, part 4, 528—542 (1960).

REVZIN, A. M., and H. KARTEN: Rostral projections of the optic tectum and the nucleus rotundus in the pigeon. Brain Res. **3**, 264—276 (1967).

SCHARRER, E.: Functional organization of the brain. In: Neuropharmacology (H. A. ABRAMSON, ed.), p. 90—106. New York, N.Y.: Josiah Macy, Jr. Foundation 1955.

—, and J. SINDEN: A contribution to the "chemoarchitectonics" of the optic tectum of the brain of the pigeon. J. comp. Neurol. **91**, 331—336 (1949).

SHANKLIN, W. M.: The central nervous system of *Chamaeleon vulgaris*. Acta zool. (Stockh.) **11**, 425—490 (1930).

SHEN, S., P. GREENFIELD, and E. J. BOELL: The distribution of cholinesterase in the frog brain. J. comp. Neurol. **102**, 717—743 (1955).

SHIMIZU, N., and N. MORIKAWA: Histochemical studies of succinic dehydrogenase of the brain of mice, rats, guinea pigs and rabbits. J. Histochem. Cytochem. **5**, 334—345 (1957).

— —, and Y. ISHII: Histochemical studies of succinic dehydrogenase and cytochrome oxidase of the rabbit brain, with special reference to the results in the paraventricular structures. J. comp. Neurol. **108**, 1—21 (1957).

SINDEN, J., and E. SCHARRER: Distribution of certain enzymes in the brain of the pigeon. Proc. Soc. exp. Biol. (N.Y.) **72**, 60—62 (1949).

Part II. Basal Structures of the Brainstem in Reptiles and Bird

Introduction

Evolutionary relationships between structures in the submammalian brain and the brain of mammals often have eluded conclusive analysis in investigations based on morphological methodology. In addition to the application of modern neurophysiological and biochemical techniques, the employment of histochemical methods offers a further means of functional analysis and comparison of brain structures in vertebrates at different phylogenetic levels. The latter has the advantage over much of the usual biochemical technique in its ability to localize chemical entities to particular cell types and neuropil units within a heterogeneous field, permitting the establishment of a "chemoarchitectonics" (SCHARRER, 1955) in parallel to classical cytoarchitectonics. It is to be anticipated that continued developments in histochemical techniques will bring rich rewards in functional neurological analysis. This is notably foreshadowed in the observations recently made by the Swedish teams of investigators, using FALCK's (1962) fluorescence method for the demonstration of monoamines (e.g. DAHLSTRÖM and FUXE, 1965a, b; FALCK, 1964).

The observations presented here were limited to the activities of several enzymes which are common to all tissues, rather than ones especially significant in brain. There was, however, the advantage that many studies of the distribution and intensity of these activities in various mammalian brains had been published (comprehensively reviewed by FRIEDE, 1966), to which reference could be made. This increased the likelihood of generality in the comparisons made here between mammal and submammal over those which might have been made if only the observations on the one mammal included in this study had been available.

In Part I of this study (BAKER-COHEN, 1968), histochemical characteristics of striatal structures in the brains of several reptiles and a bird were described. An attempt was made to relate these findings to observations on the basal ganglia of mammals, both with respect to histochemical characteristics and to published morphological and physiological data. Since the non-cortical connections of the striatum are through the lateral forebrain bundle, it seemed logical to follow this great fiber tract into the brain stem, and consider structures related to it. Hence, many of the structures to be described in the present report will be nuclei closely associated functionally with the lateral forebrain bundle. Among these are three which have been believed to represent forerunners, in whole or part, of the mammalian entopeduncular and subthalamic nuclei, and the substantia nigra.

Materials and Methods

1. Animals

The specimens studied were the same ones enumerated in Part I of this study (BAKER-COHEN, 1968). The reptilian types included individuals of: the turtle species, *Pseudomys floridana*, *Chrysemys picta* and *Testudo graeca;* the lizard species, *Anolis carolinensis* and *Phrynosoma cornutum;* and the young spectacled caiman, *Caiman sclerops.* The bird was represented by the shell parakeet, *Melopsittacus undulatus*, and the mammal by the laboratory mouse.

2. Methods

Part I of this study should be consulted for details and literature references on the histochemical methods. Fresh frozen cryostat sections of brains were incubated for four enzymes: succinate dehydrogenase (SDH), nicotinamide-adenine dinucleotide tetrazolium reductase (NAD diaphorase), acid phosphatase (AcPase) and thiamine pyrophosphatase (TPPase) Calcium-formol-fixed frozen sections were incubated for acid phosphatase.

Interrupted serial sections were usually made of the larger brains of turtle, caiman, parakeet and mouse. Complete serial sections generally were made of the small lizard brains. In either case, sets of adjacent sections usually were incubated for three or four different enzymes, so that detailed comparisons of enzyme distributions could be made.

Observations

As in Part I of this study, each of the structures to be discussed will be introduced briefly, so far as possible, from the comparative neurological literature. It is felt that this procedure is necessary for the sake of clarity, in view of the varied, and sometimes conflicting, terminologies applied to submammalian brain structures by different investigators. Structures in the turtle brain generally will be taken up first, followed by observations and comparisons of corresponding ones in the caiman and lizard brains. A short discussion of suggested mammalian homologies often will be necessary as a prelude to description of the appropriate parts of the mouse brain. The parakeet brain will be brought into the present report only in connection with structures in the midbrain tegmentum.

A. Ventral Thalamus and Subthalamus

1. Nucleus of the Dorsal Supraoptic Decussation

In the present histochemical preparations of the ventral thalamus of turtles, a group of large multipolar neurons intercalated in the dorsal peduncle of the lateral forebrain bundle was especially prominent. Preliminary study suggested that these cells lay within the nucleus rotundus component of the dorsal peduncle (tractus thalamo-striatalis intermedius, ARIENS KAPPERS *et al.*, 1960). However, in further study of preparations sectioned in three planes, it was concluded that their position was in the base of the undivided dorsal peduncle. This group of neurons clearly represented the nucleus identified as the nucleus of the dorsal supraoptic decussation by PAPEZ (1935) in turtles, and also seemed to correspond with the "nucleus entopeduncularis" shown in *Testudo graeca* by DE LANGE (1913, Fig. 32) and in *Sphenodon* by DURWARD (1930).

The large neurons and cell processes of this nucleus in the turtle were strikingly stained for SDH and NAD diaphorase (Figs. 1—4), and also were prominent in AcPase preparations (Figs. 5 and 28). In the dehydrogenase preparations, the scanty neuropil of this nucleus was deeply stained. The cells resembled those of the globus pallidus, but were more strongly stained, especially for AcPase (Fig. 28).

In close topographical relation to the above-described nucleus, and at first, from the examination of transverse sections, thought to be part of it, lay another

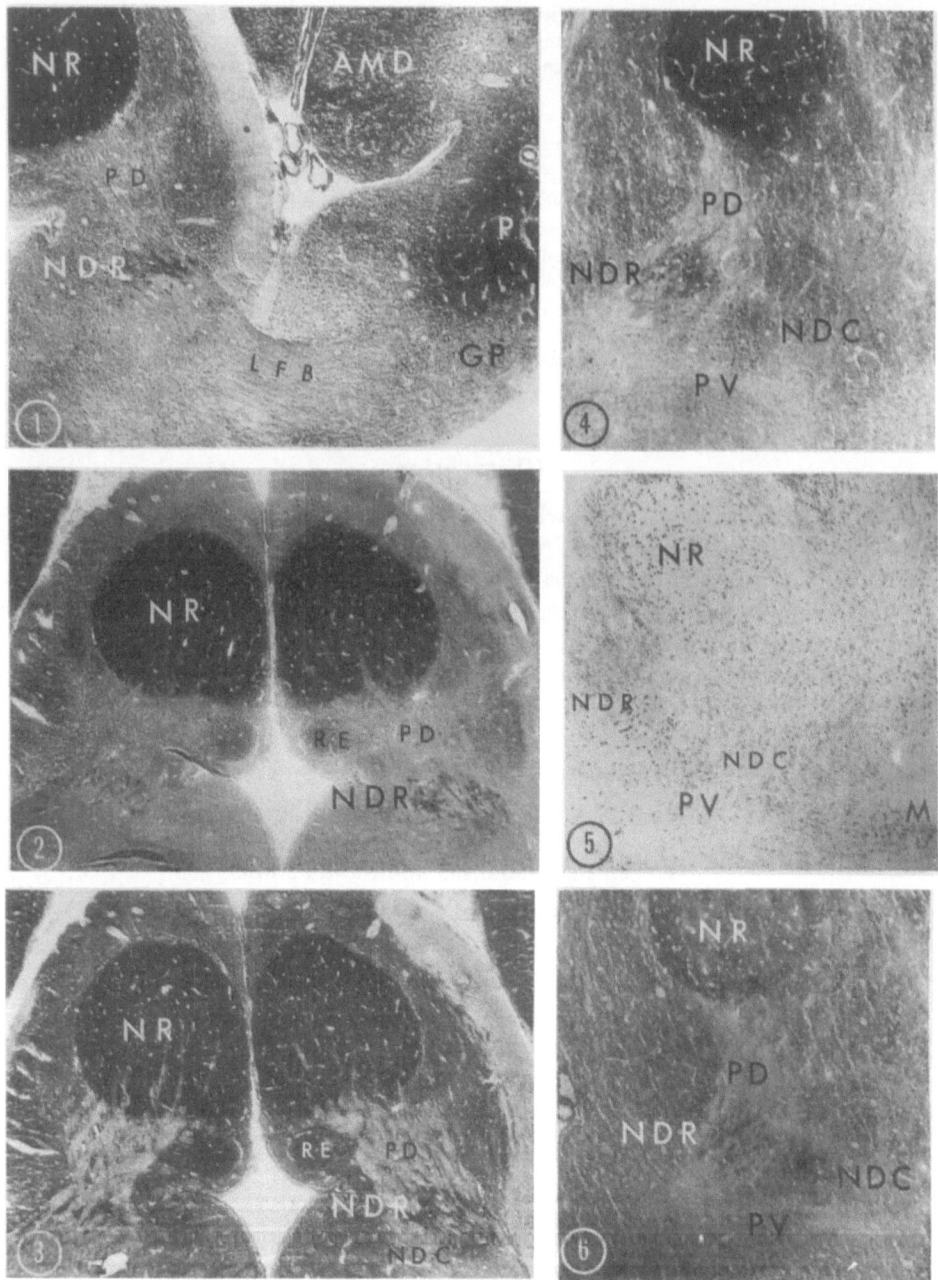

Fig. 1—6

Fig. 1. Eastern Painted Turtle. Coronal section, SDH. The passage of the lateral forebrain
bundle between hemisphere and thalamus is seen here, as are its relations to structures in
both areas. At left in the thalamus, the strongly stained neurons of the nucleus of the dorsal
supraoptic decussation, rostral, appear in relation to the dorsal peduncle, below the intensely
stained nucleus rotundus. At right in the hemisphere, dark neurons of the globus pallidus
appear among the fibers of the lateral forebrain bundle, adjacent to the intensely stained
palaeostriatum (= putamen of JOHNSTON, 1923). ×30

Abbrevations for Figures [1]

AMD	Antero-medial part of dorsal ventricular ridge (turtle)	CI	Inferior colliculus
C	Cerebellum	DA	Nucleus dorsolateralis anterior thalami
CA	Anterior commissure	DLA	Dorsolateral area (striatum)
DP	Nucleus dorsalis commissure posterior	NST	Nucleus tractus strio-tegmentalis
		NX	"Nucleus X"
DS	Supraoptic decussations	N III	Oculomotor nucleus
E	Entopeduncular nucleus	O	Optic tract
F	Fornix	OC	Optic chiasma
FM	Nucleus interstitialis fasciculus medialis	OT	Optic tectum
		OV	Nucleus ovalis
FR	Fasciculus retroflexus	P	Putamen
GL	Lateral geniculate nucleus	PD	Dorsal peduncle of the lateral forebrain bundle
GP	Globus pallidus		
H	Habenula	PN	Pretectal nucleus
HF	Hippocampal formation	PV	Ventral peduncle of the lateral forebrain bundle
HT	Hypothalamus		
IP	Interpeduncular nucleus	R	Red nucleus
IS	Nucleus isthmi, caudal part	RE	Nucleus reuniens
ISR	Nucleus isthmi, rostral part	RT	Nucleus reticularis thalami
LFB	Lateral forebrain bundle	S	Septal region
M	Mammillary nucleus	SNC	Substantia nigra zona compacta
NDC	Nucleus of the dorsal supraoptic decussation, caudal	SNL	Substantia nigra pars lateralis
		SNR	Substantia nigra zona reticulata
NDR	Nucleus of the dorsal supraoptic decussation, rostral	ST	Subthalamic nucleus
		TH	Thalamus
NPM	Nucleus profundus mesencephali	V III	Third ventricle
NR	Nucleus rotundus thalami	III	Oculomotor nerve

Fig. 2. Eastern Painted Turtle. Coronal section, SDH. The strongly stained neurons and scanty neuropil of the nucleus of the dorsal supraoptic decussation, rostral, appear among fibers of the lateral forebrain bundle below the nucleus rotundus. The moderately stained oval nuclei adjacent to the ventricle below nucleus rotundus are the nuclei reuniens. ×25

Fig. 3. Eastern Painted Turtle. Section adjacent to that shown in Fig. 2. NAD diaphorase. The nucleus of the dorsal supraoptic decussation, rostral, and the nucleus rotundus are highly active for this enzyme, as well as for SDH. Part of the nucleus of the dorsal supraoptic decussation, caudal, with lighter staining begins to show below NDR. The relatively stronger staining in other parts of the dorsal thalamus causes the dorsal peduncle of the lateral forebrain bundle to stand out more sharply than it does in the SDH preparation. ×25

Fig. 4. Western Painted Turtle. Sagittal section, SDH. The more prominent nucleus of the dorsal supraoptic decussation, rostral, appears among fibers of the dorsal peduncle, and the less prominent nucleus of the dorsal supraoptic decussation, caudal, lies in the angle between the dorsal and ventral peduncles, or partly in the ventral peduncle. ×35

Fig. 5. Western Painted Turtle. Section adjacent to that shown in Fig. 4. AcPase. Neurons of the nucleus rotundus and the nucleus of the dorsal supraoptic decussatis, rostral, show especially strong activity. The cells of the nucleus of the dorsal supraoptic decussation, caudal, are less strongly stained. The cluster of dark cells at the lower right form part of the mammillary nuclei. ×27.5

Fig. 6. Western Painted Turtle. Section close to those shown in Fig. 4 and 5. NAD adiphorase. The relative activities seen in the rostral and caudal nuclei of the dorsal supraoptic decussation are similar for this enzyme to those shown for SDH. The light staining of the nucleus rotundus seen here is an artifact of this particular section. ×35

---

[1] In all figures, representations of sagittal sections are oriented so that the rostral pole lies to the left. Horizontal sections are oriented so that the rostral pole lies at the top of the figure.

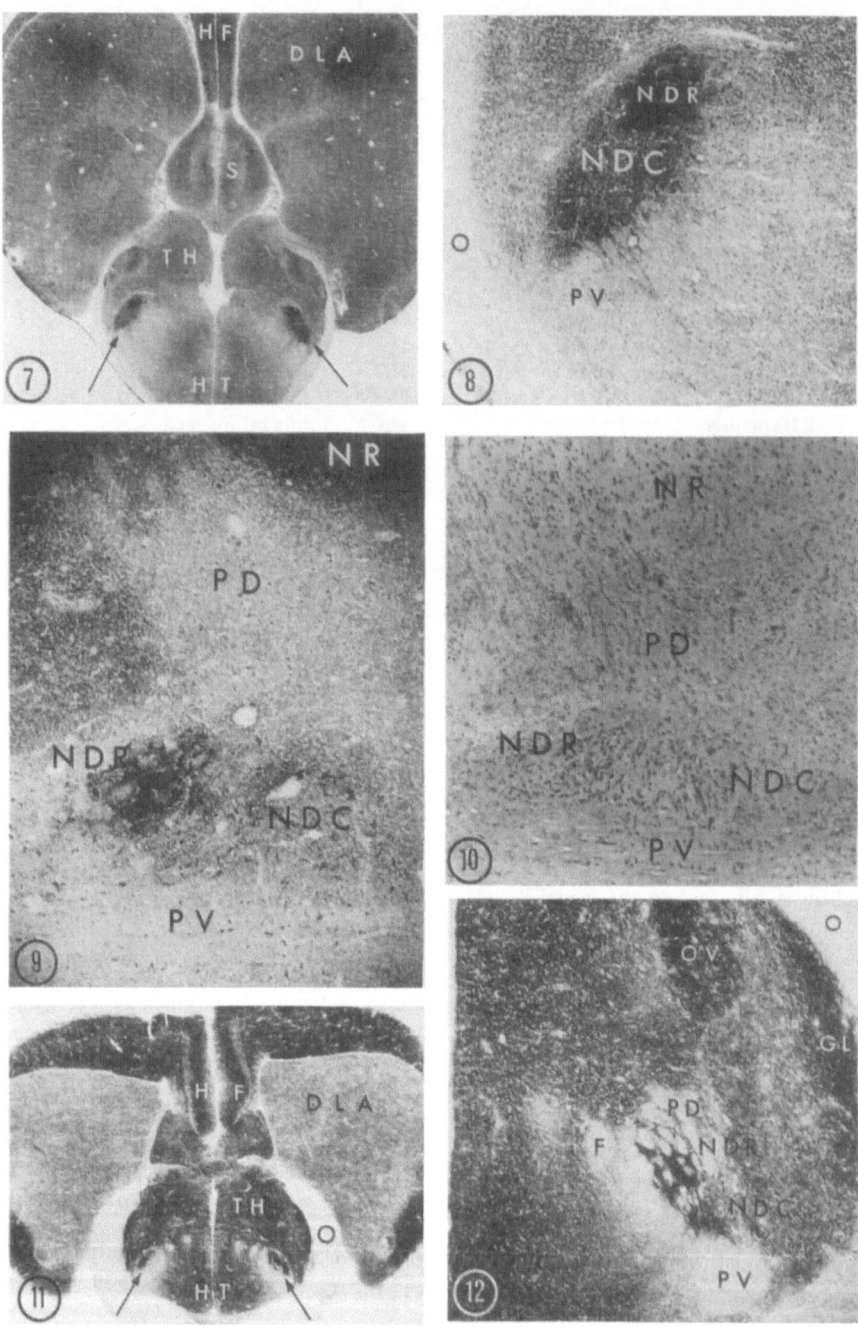

Fig. 7—12

Fig. 7. Caiman. Coronal section, SDH. Two nuclei showing prominent SDH activity occur in close physical relation to one another, and to the lateral forebrain bundle, in the rostral thalamus (arrows). The very deeply stained dorsomedial nucleus is here called "nucleus of the dorsal supraoptic decussation, rostral", although it may possibly represent a part of the interstitial nucleus of the olfactory projection tract. The more moderately stained ventro-lateral nucleus is labelled as the "nucleus of the dorsal supraoptic decussation, caudal". A higher power view of these is seen in Fig. 8. ×8

group of neurons, many of which stained quite strongly for SDH, NAD diaphorase, and to a lesser degree, AcPase. These cells were embedded in a mass of neuropil which showed moderate activity for SDH and NAD diaphorase (Figs. 4, 6). Study of sagittal and horizontal sections revealed that this nucleus was separate from that first described, but overlapped it rostro-caudally. From sagittal sections, it appeared that the second nucleus lay in a angle between components of the dorsal peduncle and the ventral peduncle of the lateral forebrain bundle, and seemingly partly among fiber bundles of the ventral peduncle (Figs. 4, 6). This second nucleus does not appear to have been differentiated by PAPEZ, and may have been included as part of his nucleus of the dorsal supraoptic decussation. For lack of more definitive identification the two nuclei described will be referred to in further discussion and labelling of figures as "nucleus of the dorsal supraoptic decussation, rostral, and caudal".

In the alligator, HUBER and CROSBY (1926) noted that the nucleus of the dorsal supraoptic decussation lies in the angle between the anterior thalamo-striatal tract and the remainder of the dorsal components of the lateral forebrain bundle. The interpretation of this area in the histochemical preparations of the young caiman brain was complicated by the presence of two contiguous neuron groups with prominent, but differing, enzymatic activities, both of which appeared to be embedded in the dorsal peduncle of the lateral forebrain bundle. In trans-verse sections, a medial rather narrow cell group was embedded in a neuropil mass which stained in part so heavily for SDH and NAD diaphorase that the neuron bodies scarcely could be distinguished. Ventro-lateral and closely applied to this lay a larger collection of neurons which also stained strongly for SDH and

Fig. 8. Caiman. Detail of section shown in Fig. 7. Neuropil of the "nucleus of the dorsal supraoptic decussation, rostral" is so intensely stained that the cell bodies cannot be distin-guished. Neurons visible in the more moderately stained neuropil of the "nucleus of the dorsal supraoptic decussation, caudal" show strong activity. Fiber bundles are visible extending from the latter nucleus into the medioventral hypothalamic region. ×45

Fig. 9. Caiman. Sagittal section, SDH. The very intensely active "nucleus of the dorsal supra-optic decussation, rostral", appears in this section to lie among fibers of the dorsal peduncle of the lateral forebrain bundle, and the more lightly stained "nucleus of the dorsal supra-optic decussation, caudal" appears to be largely in the area of separation of dorsal and ventral peduncles. These relations seem to correspond to those seen in the turtle. Compare with Fig. 4 to 6. ×60

Fig. 10. Caiman. Section close to that shown in Fig. 9. AcPase. Neurons of the "nucleus of the dorsal supraoptic decussation, caudal" show less activity for AcPase than those of the "nucleus of the dorsal supraoptic decussation, rostral". The scattered neurons of nucleus rotundus show strong activity. ×40

Fig. 11. Anolis. Coronal section, SDH. The strongly stained nucleus of the dorsal supraoptic decussation, rostral, lies among the fibers of the dorsal portion of the lateral forebrain bundle (arrows). The latter in this section is beginning to separate into dorsal and ventral peduncles, with the nucleus of the dorsal supraoptic decussation, caudal, appearing in the separation. A higher power view of a section close to this one is shown in Fig. 12. ×20

Fig. 12. Anolis. Section close to that shown in Fig. 11. SDH. The intensely stained reticularly arranged nucleus of the dorsal supraoptic decussation, rostral, appears among fascicles of the dorsal peduncle of the lateral forebrain bundle, and the less strongly stained, more homogeneously structured nucleus of the dorsal supraoptic decussation, caudal, lies in the separation between dorsal and ventral peduncles. The nucleus ovalis and lateral geniculate nucleus show prominent activity in neuropil. ×80

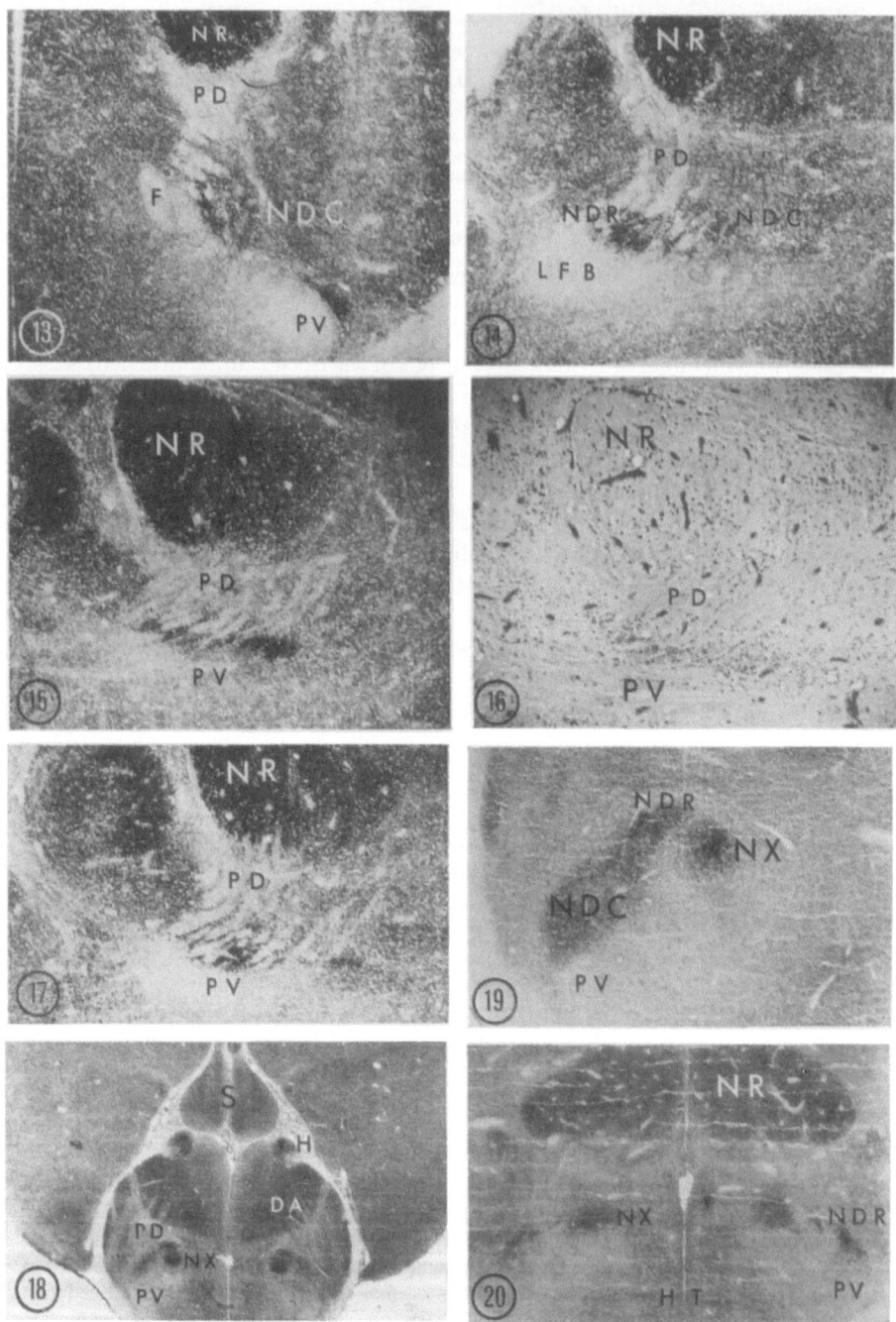

Fig. 13—20

Fig. 13. *Anolis*. Coronal section, SDH. Section caudal to that shown in Fig. 12, as the nucleus rotundus makes its appearance. The deeply stained nucleus of the dorsal supraoptic decussation, rostral, is disappearing and the less stained nucleus of the dorsal supraoptic decussation, caudal, has enlarged in the deepening space between the dorsal and ventral peduncles.

The oval fiber tract just to the left of the dorsal peduncle is the fornix. ×80

NAD diaphorase, but were embedded in a mass of neuropil which stained only moderately for the two dehydrogenases (Figs. 7—9). Neurons of the first group showed very strong AcPase activity, with less staining of the cells in the second group (Fig. 10). In those sagittal sections, in which both nuclei were in view, the relative histochemical activities and the positional relations of these two nuclei in the caiman brain appeared similar to those of the two nuclei described above in the turtle (compare Figs. 4—6 and 9—10). They differed in that the cells and neuropil of the more rostrally appearing nucleus in the caiman formed a solid mass of relatively small extent, but was loosely reticular in the turtle.

Study of the transverse and sagittal sections, and comparison with HUBER and CROSBY's figures, indicated that the position of the more caudal and lateral nucleus in the caiman preparations corresponded to that of the nucleus of the dorsal supraoptic decussation in the alligator. In SDH preparations, from the neuropil of the lateral nucleus fiber bundles could be seen to emanate and extend ventro-medially past the ventral peduncle towards the basal part of the hypothalamus

Fig. 14. *Anolis*. Sagittal section, SDH. The intercalation of the strongly stained nucleus of the dorsal supraoptic decussation, rostral, among components of the dorsal peduncle of the lateral forebrain bundle is clearly seen here and in Fig. 17. The less enzymatically active nucleus of the dorsal supraoptic decussation, caudal, appears, rather indistinctly, in the angle between dorsal and ventral peduncular components. Compare with nuclei in the turtle in Fig. 4, and in caiman in Fig. 9. ×80

Fig. 15. *Anolis*. Sagittal section, SDH. The elongated distribution of the nucleus of the dorsal supraoptic decussation, rostral, in relation to more medial components of the thalamic radiations is illustrated here. This rather extensive structure can also be seen in a horizontal section in Fig. 25. ×80

Fig. 16. *Anolis*. Section adjacent to that shown in Fig. 15. AcPase. Although small at this magnification, neurons of the nucleus of the dorsal supraoptic decussation, rostral, can be seen to show relatively strong AcPase activity (compare with Fig. 15). The scattered neurons of nucleus rotundus also show relatively strong staining. The nucleus rotundus is partially outlined by darkened blood vessels, whose strong staining in AcPase preparations of *Anolis* brains interferes with observation of the cytological picture. ×80

Fig. 17. *Anolis*. Sagittal section, SDH. A section somewhat medial to that shown in Fig. 14, to emphasize the relation of the nucleus of the dorsal supraoptic decussation, rostral, to fascicles of the dorsal peduncle. ×80

Fig. 18. Caiman. Coronal section, SDH. At this level of the rostral thalamus a spherical nucleus showing strong SDH activity appears just below the "nucleus of the dorsal supraoptic decussation, rostral", and medial to the "nucleus of the dorsal supraoptic decussation, caudal". This is referred to here as "nucleus X". Above these nuclei, fiber bundles of the dorsal peduncle are seen cleaving the strongly stained, arrowhead-shaped, nucleus dorso-lateralis anterior. ×8

Fig. 19. Caiman. Coronal section, SDH. A section close to that shown in Fig. 8, to show the relations of "nucleus X" to the "nucleus of the dorsal supraoptic decussation — rostral and caudal". ×30

Fig. 20. *Phrynosoma*. Coronal section, SDH. Strong enzymatic activity is seen in neurons and neuropil of the nucleus of the dorsal supraoptic decussation, rostral, which lies laterally in relation to the lateral forebrain bundle. Even stronger staining appears in the rounded nucleus which appears to correspond to that in the caiman here called "nucleus X". This latter nucleus lies just dorsal to the medial forebrain bundle. The heaviliy stained nucleus rotundus appears above. ×37.5

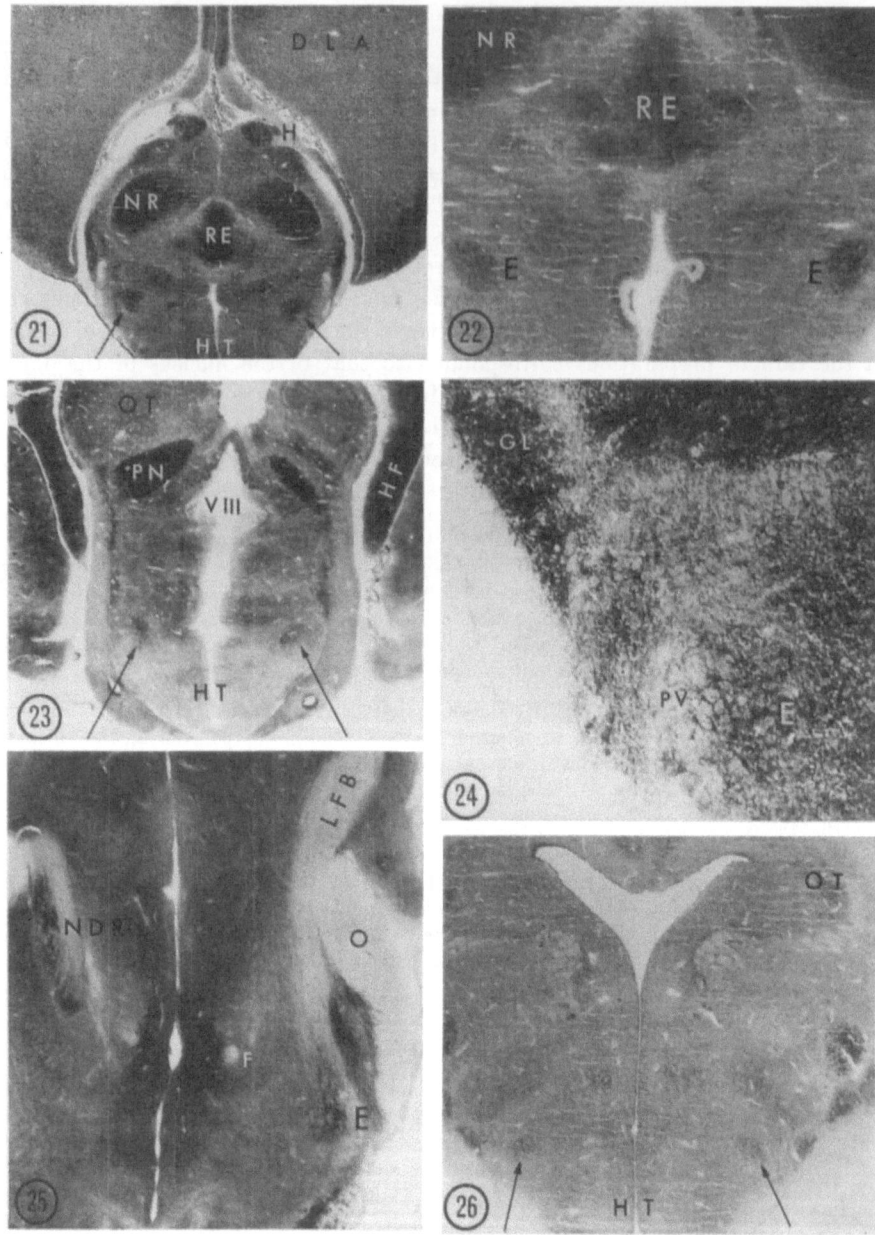

Fig. 21—26

Fig. 21. Caiman. Coronal section SDH. The entopeduncular nuclei (arrows) appear in the ventral peduncles of the lateral forebrain bundle in sections through the midthalamus. The neurons of this nucleus contain strong dehydrogenase activity and that in its neuropil is moderately strong (see also Figs. 22 and 31). The nucleus rotundus and nucleus reuniens show the most intense dehydrogenase activity in the thalamus. ×8

Fig. 22. Caiman. Coronal section, SDH. In this section, slightly caudal to that shown in Fig. 21, the entopeduncular nuclei may be seen to have strongly stained cell bodies embedded in a moderately active neuropil. The pockets at the sides of the ventricle represent the "vascular organ" characteristic of reptile brains (ARIENS KAPPERS et al., 1960). ×20

(Fig. 8). This also suggested that the caudal neuron group indeed represented the nucleus of the dorsal supraoptic decussation. Comparison with HUBER and CROSBY's Fig. 4 and 5 suggested that the deeply stained rostro-medial nucleus might represent part of the interstitial nucleus of the olfactory projection tract, although as a histochemically distinctive group it did not extend into the base of the hemisphere, as described by HUBER and CROSBY and illustrated in their Fig. 3. Since the histochemical preparations often did not differentiate cohering fiber bundles well, it was possible that the olfactory projection tract at the level of the deeply stained nucleus might appear to be part of the dorsal peduncle. However, for the sake of simplification, these two nuclei in the caiman brain will be labelled in the same manner as the two similar appearing nuclei in the turtle brain.

Analysis of the dorsal peduncular area of the caiman brain in the histochemical preparations was further complicated by the presence of a third very prominent nucleus closely adjacent to the two just described. Medial to the probable nucleus of the dorsal supraoptic decussation, and just below the possible interstitial nucleus of the olfactory projection tract, appeared a cell group whose neurons stained heavily for SDH. These were embedded in a spherical mass of neuropil which also stained very strongly for SDH. This nucleus is shown in Figs. 18 and 19 and is here referred to as "nucleus X". The cells were strongly stained for AcPase, and appeared similar in shape and staining to the possible interstitial nucleus of the olfactory projection tract. This nucleus may appear without labelling in HUBER and CROSBY's Fig. 5.

In *Anolis*, two nuclei closely resembling those described in the turtle appeared in the histochemical preparations and will be given the same appellations. The more anterior one was loosely organized and contained extremely strong concentrations of SDH activity in its neurons and scanty neuropil (Figs. 11—14). As in the turtle, it was also very prominent in the AcPase preparations. The more caudal nucleus was less well-defined than in the turtle, and the homogeneous mass of neuropil delineated by SDH staining seemed to be relatively small. The cells were

Fig. 23. Eastern Painted Turtle. Coronal section, SDH. The entopeduncular nuclei (arrows) in the ventral thalamus show moderately strong enzymatic activity in cells and neuropil. Fibers of the supraoptic decussations are visible below in the poorly stained hypothalamus (by virtue of brownish discoloration, not blue enzymatic reaction). The heavily stained structures beneath the optic lobes are components of the pretectal nuclear grouping. ×11.5

Fig. 24. *Anolis*. Coronal section, SDH. Neurons and processes of the entopeduncular nucleus, embedded among fibers of the ventral peduncle, are strongly stained. The scanty neuropil shows moderate activity. ×80

Fig. 25. *Anolis*. Horizontal section, SDH. At the right, the lateral forebrain bundle is seen extending between the hemisphere (out of sight, above) into the ventral thalamic region. The entopeduncular nucleus, lying among its fibers, shows quite strong activity. At the left, part of the lateral forebrain bundle at a more dorsal and rostral level (owing to obliquity in cutting the section) has associated with it the elongated, deeply stained nucleus of the dorsal supraoptic decussation, rostral. The small round unstained fiber tract at either side of the midline is the fornix. ×37.5

Fig. 26. *Phrynosoma*. Coronal section, SDH. The entopeduncular nuclei (arrows) show similar enzymatic characteristics to their counterparts in the other reptile species. Even in this under-incubated preparation the neurons are dark and the nucleus as a whole stands out moderately well. ×25

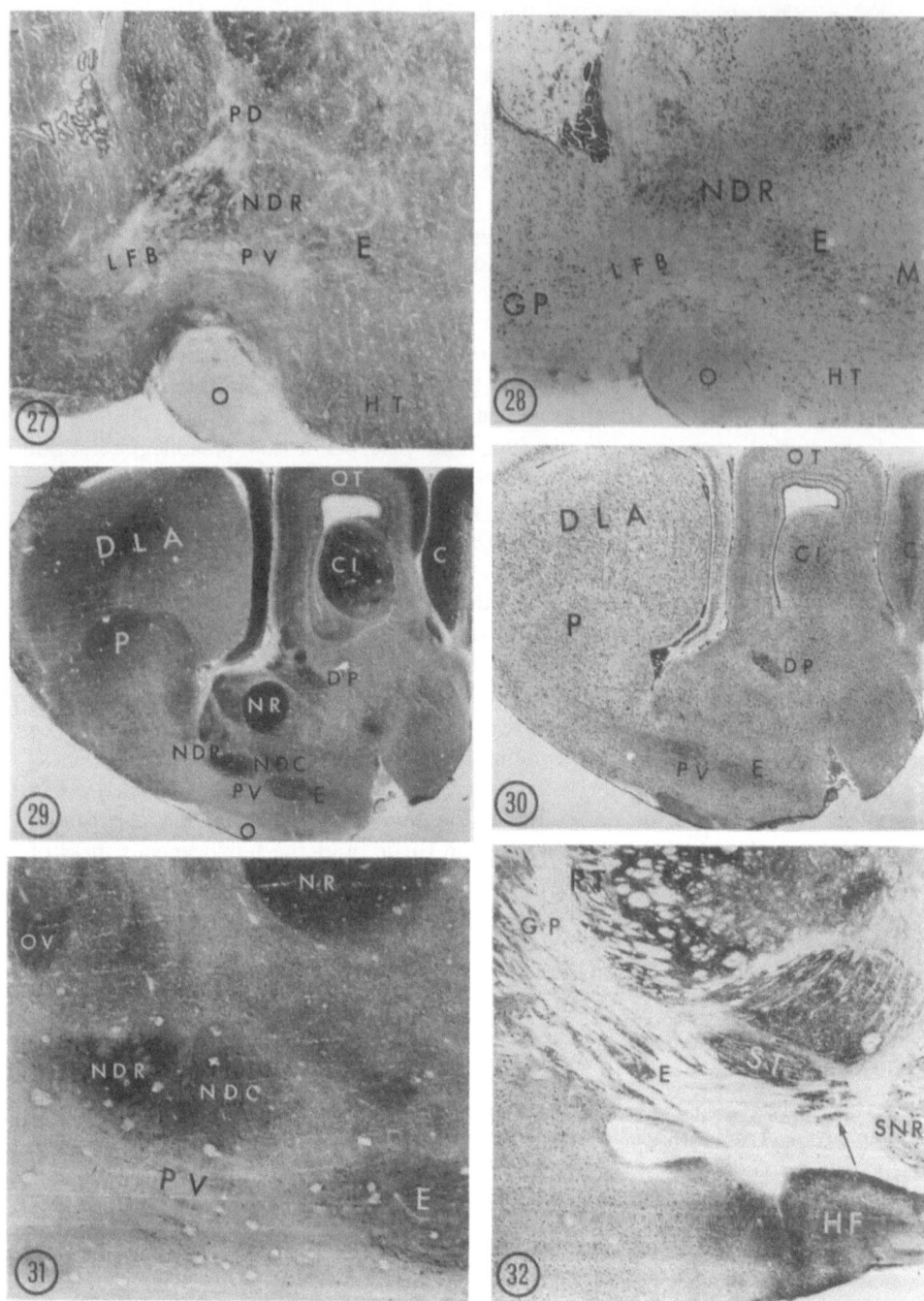

Fig. 27—32

Fig. 27. Western Painted Turtle. Sagittal section, NAD diaphorase. Part of the entopeduncular nucleus appears in its relation to the ventral peduncle of the lateral forebrain bundle, some distance caudal to the nuclei of the dorsal supraoptic decussation. Compare with the adjacent section shown in Fig. 28. ×27.5

weakly or moderately stained for AcPase. The relations of both nuclei in *Anolis* to components of the lateral forebrain bundle appeared to be similar to those of the two nuclei described in the turtle.

In *Anolis*, the olfactory projection tract and its interstitial nucleus was somewhat separated from the dorsal peduncle, and the tract could be followed over the peduncle to the basal part of the hemisphere. The interstitial neurons and their prominent processes were strongly stained for SDH, but were considerably smaller than the deeply stained cells associated with the dorsal peduncle.

In *Phrynosoma*, a loosely reticular nucleus corresponding to the more anterior and medial one of the turtle and *Anolis* ("nucleus of the dorsal supraoptic decussation, rostral") appeared in the dorsal part of the lateral forebrain bundle. Like these, its staining for SDH and AcPase was highly prominent. A second nucleus showing less SDH staining and more compact neuropil was not apparent. Rather, it seemed that a more lightly stained, but still reticular continuation of the first nucleus extended caudally and ventro-laterally into the area of divergence of the ventral peduncle from the dorsal components. Fiber bundles could be seen extending from the supraoptic decussations in the direction of these cells.

In *Phrynosoma*, a nucleus somewhat similar in histochemical activity and position to that designated "nucleus X" in the caiman appeared just above the medial forebrain bundle (Fig. 20). It was made up of a rounded mass of neuropil

Fig. 28. Western Painted Turtle. Section adjacent to that shown in Fig. 27. AcPase. The most strongly stained neurons visible here are those of the nucleus of the dorsal supraoptic decussation, rostral. Less prominent but still highly active are neurons of the globus pallidus, entopeduncular nucleus and mammillary nucleus. ×27.5

Fig. 29. Caiman. Sagittal section, SDH. The topographical relations of the nuclei of the dorsal supraoptic decussation and the entopeduncular nucleus are similar to those of the nuclei so identified in the turtle. However, all of these nuclei in the caiman show a more compact and clearly defined mass of neuropil than those in the turtle. Other structures showing intense SDH activity which appear in this section include the hippocampal cortex, inferior colliculus and cerebellar cortex. ×8

Fig. 30. Caiman. Section adjacent to that shown in Fig. 29. AcPase. Relatively strong activity is seen in neurons of the nuclei of the dorsal supraoptic decussation and the entopeduncular nucleus. The very intensely stained cluster of cells above in the pretectal area is the nucleus dorsalis commissure posterior. ×8

Fig. 31. Caiman. Higher power view of the section shown in Fig. 29. The strong staining of cells in the entopeduncular nucleus is clearly seen here. The appearance of the nucleus is very similar to that of the subthalamic nucleus of the mouse, as shown in Fig. 32. Rostrally, the very intense enzymatic activity and compact structure of the "nucleus of the dorsal supraoptic decussation, rostral" contrasts with the lesser activity in the "nucleus of the dorsal supraoptic decussation, caudal". The relations of these two structures to the dorsal peduncle of the lateral forebrain bundle is not clearly seen in this section. ×40

Fig. 32. Mouse. Sagittal section, SDH. Although the neurons of the globus pallidus, entopeduncular nucleus and reticular nucleus of the thalamus all contain about equally strong SDH activity, the neuropil of the latter two nuclei shows deeper staining than that of the globus pallidus. The subthalamic nucleus shows very strong activity in both neuropil and cell bodies. The dark area above the subthalamic nucleus is the zona incerta. The substantia nigra zona reticulata, like the globus pallidus, shows deeply stained neurons in rather weakly active neuropil. In the peduncle between the subthalamic nucleus and the substantia nigra is a clump of neurons and neuropil which appears to be a bit of the entopeduncular nucleus that has "migrated" in the direction of the substantia nigra (arrow). ×30

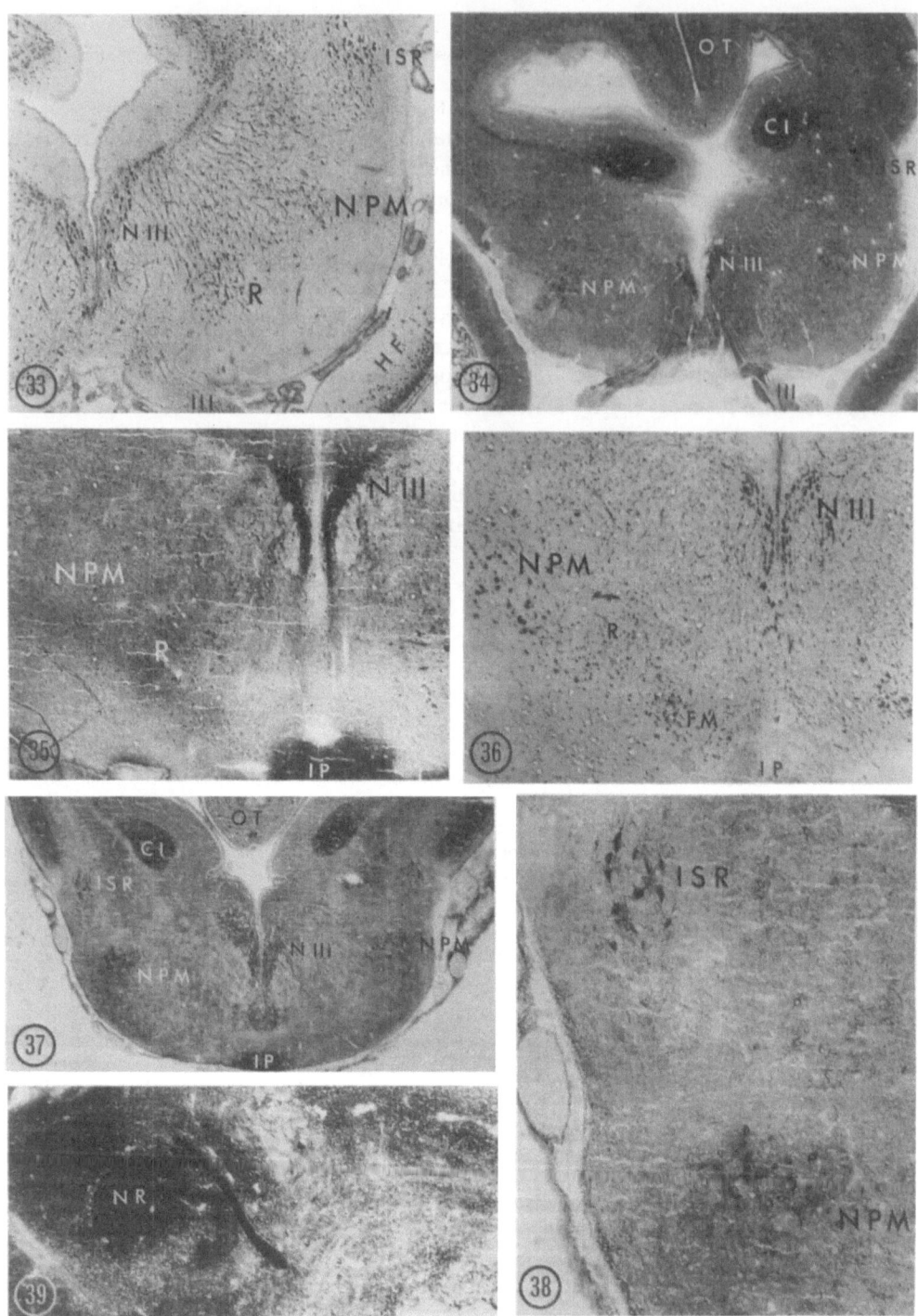

Fig. 33—39

having strong SDH activity. The cells were moderately stained for AcPase, but did not stand out over the neuropil background for SDH.

HUBER and CROSBY (1933) considered the nucleus of the dorsal supraoptic decussation in reptiles to be homologous with part of the mammalian entopeduncular nucleus. PAPEZ (1935) also made this suggestion, perhaps more tentatively, noting that the entopeduncular nucleus of mammals "appears to be related to a part of the supraoptic decussations". The latter relation had been detailed in carnivora by RIOCH (1929). The entopeduncular nucleus of lower mammals has been demonstrated to be homologous with the medial segment of the globus pallidus in primates (NAUTA and MEHLER, 1961), and lies in the ventral peduncle of the lateral forebrain bundle. In the latter respect it differs from the nucleus of the dorsal supraoptic decussations as described in the literature on reptilian brains, which is interstitial in the "principal thalamic radiation" (PAPEZ, 1935). Because of their common location, and certain histochemical similarities, the nuclei described above in the reptile brains therefore will be discussed in comparison with the mammalian reticular nucleus of the thalamus, as well as with the mammalian entopeduncular nucleus.

Fig. 33. Eastern Painted Turtle. Coronal section, AcPase. The most intensely stained neurons appearing in this section are those of the oculomotor nucleus and the rostral part of the nucleus isthmi. Cells of the red nucleus are more prominent in their AcPase activity than those of the nucleus profundus mesencephali, in contrast to their relative activities for SDH, as seen in Fig. 34. ×30

Fig. 34. Eastern Painted Turtle. Section adjacent to that shown in Fig. 33. SDH. By comparison with the adjacent section shown in Fig. 33, it is apparent that SDH activity in the red nucleus is minimal. However, strong SDH activity is seen in neurons of the nucleus profundus mesencephali. Strong staining also is seen in the cells of the rostral nucleus isthmi, and in neurons and neuropil of the oculomotor nucleus and inferior colliculus. ×20

Fig. 35. Caiman. Coronal section, SDH. In the red nucleus, neurons show strong enzymatic activity and neuropil is moderately stained. By comparison with the adjacent section shown in Fig. 36, it may be seen that rather poor activity is present in the nucleus profundus mesencephali. The neuropil of the oculomotor and interpeduncular nuclei is intensely stained. ×30

Fig. 36. Caiman. Section adjacent to that shown in Fig. 35. AcPase. Neurons of the nucleus profundus mesencephali are intensely stained, but those of the red nucleus are very little so. Neurons of the oculomotor nucleus are highly active. The small cells in the interpeduncular nucleus show little activity. The strongly stained group of cells below the red nucleus is the interstitial nucleus of the fasciculus longitudinalis medialis. ×30

Fig. 37. Western Painted Turtle. Coronal section, NAD diaphorase. Strong enzyme activity is apparent in neurons of the nucleus profundus mesencephali, oculomotor nucleus and rostral part of nucleus isthmi, and in neuropil of the interior colliculus and interpeduncular nucleus. Neuropil associated with the nucleus profundus mesencephali is moderately stained. ×20

Fig. 38. Western Painted Turtle. A detail of the section shown in Fig. 37, showing NAD diaphorase activities in the nucleus profundus mesencephali and the rostral part of nucleus isthmi. ×80

Fig. 39. Anolis. Sagittal section, SDH. The two species of lizards were unique in the possession of an habenulo-peduncular tract (fasciculus retroflexus) which was intensely stained for SDH. NAD diaphorase activity was equally strong in this tract in Anolis. Here the tract is seen curving downward behind the heavily stained nucleus rotundus. The light "cloudy" area above the fasciculus retroflexus represents fascicles of the posterior commissure. ×45

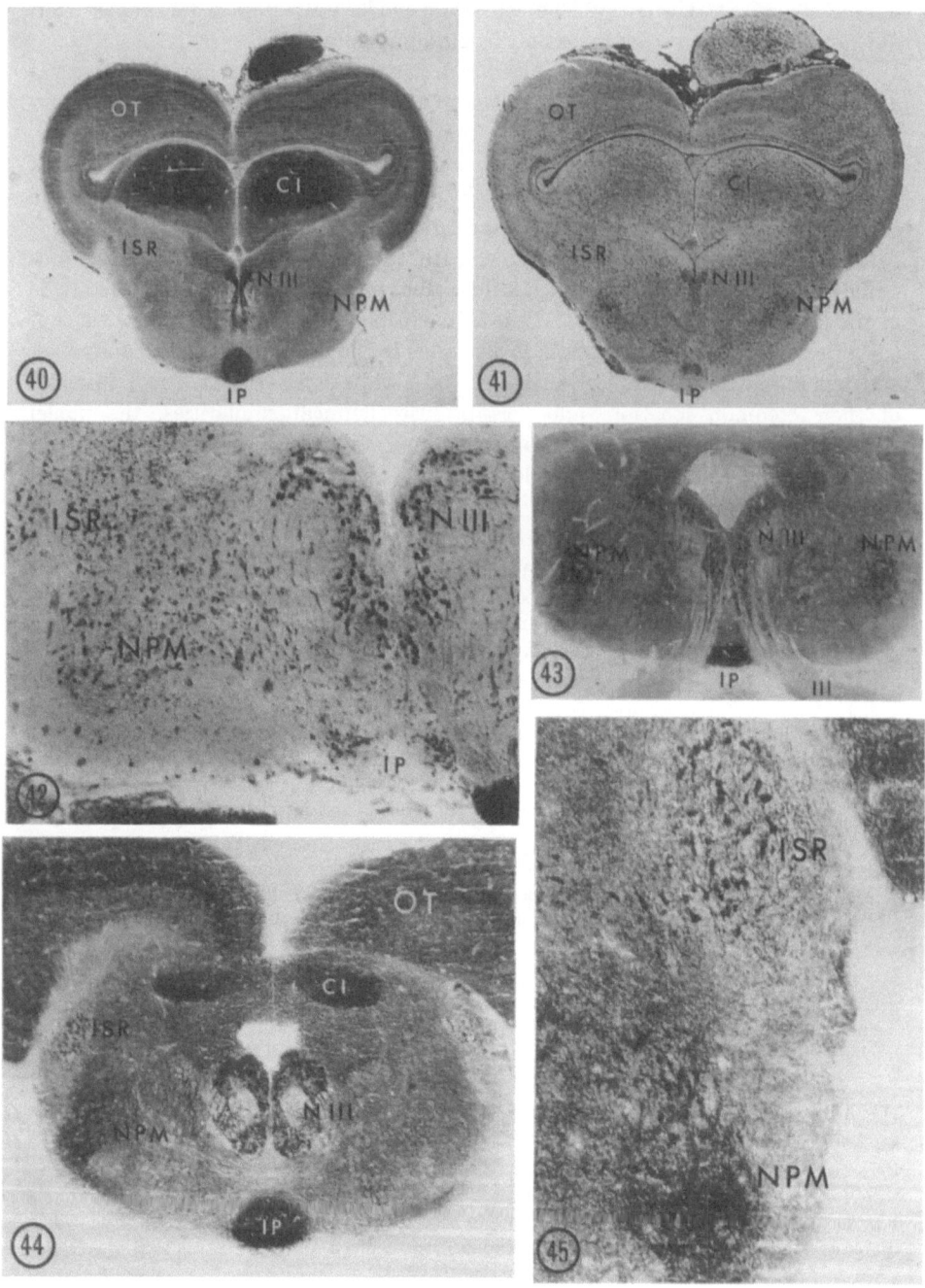

Fig. 40—45

Fig. 40. Caiman. Coronal section, SDH. Intensive enzymatic activity is seen in neuropil of the interpeduncular nucleus, oculomotor nucleus and inferior colliculus. Neurons and neuropil of the nucleus profundus mesencephali are rather weakly stained. ×8

In the present histochemical preparations of mouse brains, neurons of the thalamic reticular nucleus, the entopeduncular nucleus and the globus pallidus all were highly active for SDH and NAD diaphorase. The neuropil of the reticular nucleus was moderately to strongly stained for SDH, and that of the entopeduncular nucleus somewhat less so. Neuropil of both was considerably more active for this enzyme than that of the globus pallidus (Fig. 32), with which the entopeduncular nucleus was in partial continuity. The observations on SDH activity differed in one respect from those of FRIEDE (1961) in the guinea pig, in which he noted a "mild" reaction in neuropil of the reticular nucleus — which however, he noted to deepen rostrally. In the present AcPase preparations of the mouse brain, most neurons of the globus pallidus outdid those of both the reticular and entopeduncular nuclei in activity. Of the latter two nuclei, generally stronger AcPase staining was seen in the neurons of the reticular nucleus.

2. Entopeduncular Nucleus

The structure in reptile brains which has most generally been termed the entopeduncular nucleus lies within the ventral peduncle of the lateral forebrain bundle considerably caudal to the nucleus of the dorsal supraoptic decussation. It was figured by DE LANGE (1913, Figs. 24 and 25, but *not* Figs. 32 and 33) in *Draco volans*, by HUBER and CROSBY (1926) in *Alligator* and by PAPEZ (1935) in *Chelone midas*. At first glance, HUBER and CROSBY (1933) appeared to be suggesting a genetic relationship of this nucleus to the substantia nigra of mammals. However, their analysis was directed towards a *mesencephalic* structure which had been termed nucleus entopeduncularis by SHANKLIN (1930) (see Section B, part 2, below). The "true" entopeduncular nucleus, of the authors cited first, lies much more rostrally in the subthalamus, at the level of the pretectal nuclei and posterior commissure.

Fig. 41. Caiman. Section adjacent to that shown in Fig. 40. AcPase. Neurons of the nucleus profundus mesencephali are highly active for this enzyme. The small cells in the dorsal part of the interpeduncular nucleus appear to be strongly stained, but at the low magnification this appearance is partly due to their close aggregation. ×8

Fig. 42. *Phrynosoma*. Coronal section, AcPase. Neurons of the nucleus profundus mesencephali, as in the caiman, are intensely active for this enzyme, although in the lizard they are overshadowed by the size and activity of the neurons of the oculomotor nucleus. Small cells of the interpeduncular nucleus are quite strongly stained. ×45

Fig. 43. *Phrynosoma*. Coronal section, SDH. The nucleus profundus mesencephali and the oculomotor nucleus show strong enzyme activity in neurons and moderately strong activity in neuropil. The interpeduncular nucleus, lying between the oculomotor nerve roots, contains intense SDH staining. ×25

Fig. 44. *Anolis*. Coronal section, SDH. Neuropil associated with the nucleus profundus mesencephali is so strongly stained in this species that the neuron bodies are obscured. As in the other reptiles, neuropil of the interpeduncular and oculomotor nuclei and inferior colliculus is intensely active for SDH, as well as are neurons of the rostral part of nucleus isthmi ×31.8

Fig. 45. *Anolis*. Coronal section, SDH. Section cut slightly rostral to that shown in Fig. 44. At this magnification, strong SDH staining can be seen in neurons of the nucleus profundus mesencephali, as well as in its neuropil. Cells of the rostral nucleus isthmi, above, also show great activity. ×110

Since the entopeduncular nucleus of the caiman was better differentiated in the histochemical preparations than that of the turtle, it will be dealt with first in this case. Sagittal sections of the young caiman brain, incubated for SDH, NAD diaphorase and AcPase, showed an elongated neuron group lying in the ventral peduncle in a position corresponding with that of the entopeduncular nucleus illustrated in Huber and Crosby's (1926) sagittal section (their Fig. 17). In a transverse series, this nucleus appeared in sections at the levels of the caudal nucleus rotundus and nucleus reuniens, and of the habenulae, posterior commissure and pretectal nuclei. The neuropil of this group stained quite well for SDH, and the cell bodies strongly so, so that the nucleus stood out prominently over the general background (Figs. 21, 22, 29, 31). NAD diaphorase in the neuropil was moderate. Staining of the neurons for AcPase was weak to moderate, as compared with the cells of the nucleus of the dorsal supraoptic decussation, rostral (Fig. 30).

Sagittal sections of the turtle brain showed a corresponding nucleus which differed slightly in histochemical character from that in the caiman. Many of the neurons in this nucleus, in the turtle, were quite strongly stained for AcPase, although less so than those of the nucleus of the dorsal supraoptic decussation, rostral (Fig. 28). SDH activity in the neuropil seemed to be relatively weaker than in the caiman, and the neuronal somas were moderately active for this enzyme (Figs. 23, 27).

The entopeduncular nucleus in the lizards in general was similar to that in the turtle. The cells and their processes were strongly stained for SDH and the neuropil moderately (Figs. 24—26). NAD diaphorase staining also was pronounced. AcPase in the neuron bodies ranged from weak to moderate in *Anolis* and was strong in *Phrynosoma*.

Papez (1935) believed the entopeduncular nucleus of reptiles to be homologous with the mammalian subthalamic nucleus (corpus of Luys). This nucleus in the mouse showed strong SDH activity in its neuropil, and its neuron bodies also were deeply stained (Figs. 32, 50). This observation accords with that of Friede (1961) in the guinea pig. NAD diaphorase was relatively weaker than SDH in both cells and neuropil of the mouse nucleus, and AcPase incubation stained the cell somas strongly.

B. Midbrain Tegmentum

The reptilian nuclei which will be considered are the red nucleus, the nucleus profundus mesencephali and the interpeduncular nucleus. The parakeet brain will be brought into this section in connection with the nucleus profundus mesencephali of reptiles, and with a probable homologue of the substantia nigra of mammals.

1. Red Nucleus

The red nucleus, according to Ariens Kappers et al. (1960) is relatively well-differentiated in reptiles, and was believed to be the forerunner of the magnocellular part of this nucleus in mammals. As in mammals, it is located in the reticular formation lateral to the oculomotor nucleus. It was noted to receive a component of the ventral peduncle by Shanklin (1930) and Papez (1935).

In the histochemical preparations of reptile brains, this nucleus was most clearly differentiated in the turtle. The neuron somas and processes in this nucleus in the turtle brain, especially of the larger cells, were quite strongly stained for AcPase (more so than the still larger neurons of the nucleus profundus mesencephali — see part 2, below) (Fig. 33). SDH activity in the cells was moderate or weak, and was weak in the neuropil (Fig. 34). NAD diaphorase staining was similar to that for SDH.

In the caiman and lizards, AcPase activity in neurons of the red nucleus was generally quite weak, with the exception of some of the largest cells in the former (Fig. 36). In the lizards, as in turtles, staining for SDH was moderate in the neuron bodies and their processes and weak in the neuropil. In the caiman, however, a fair amount of SDH activity appeared in neuropil and cells of the red nucleus (Fig. 35).

In the mouse, the red nucleus was very much more prominent in the histochemical preparations than it was in the reptiles. As in the guinea pig (FRIEDE, 1959) and the cat (FRIEDE, 1961), the large neurons were intensely stained for SDH and the reticular neuropil also was highly active. NAD diaphorase was moderate in the cells and poor in the neuropil. The large neurons were highly active for AcPase and somewhat less so for TPPase.

2. Nucleus Profundus Mesencephali and Substantia Nigra

Reptiles. HUBER and CROSBY (1933), in a study of a variety of reptiles (including *Anolis*, *Phrynosoma*, turtles, snakes and *Alligator*), described the substantia nigra as a group of scattered large cells along the lateral wall of the mesencephalon at the level of the red and oculomotor nuclei. They noted that these cells received terminal fibers of the lateral forebrain bundle, and had tectal connections resembling those described for the substantia nigra of mammals. SHANKLIN (1930) had termed a similarly located large-celled nucleus in *Chameleon* the "nucleus entopeduncularis", and had noted that this nucleus formed the main terminus of the ventral peduncle. SHANKLIN's nucleus entopeduncularis was equated by HUBER and CROSBY with their substantia nigra. These nuclei described by HUBER and CROSBY and by SHANKLIN appear to represent the same cell grouping as that referred to by other students of the reptilian brain as the nucleus profundus mesencephali. The latter was figured in *Lacerta* by FREDERIKSE (1931) and in turtles by PAPEZ (1935). Such a nucleus was shown unlabelled in *Alligator* and *Testudo* by DE LANGE (1913, Figs. 50 and 51), and subsequently was labelled as nucleus profundus mesencephali by ARIENS KAPPERS (1921) in a reproduction of one of DE LANGE's figures. BECCARI's (1923, Figs. 11, 12 and 22) "sost. nera ?" also appears to represent the same cell group.

In the histochemical preparations of the turtle brain, the large neurons and cell processes of the nucleus profundus mesencephali were highly active for SDH and NAD diaphorase (Figs. 34, 37, 38). For these enzyme activities, this nucleus was much more prominent than the nearby red nucleus. Staining of neuropil for SDH was poor, but moderately strong for NAD diaphorase. In further contrast to the red nucleus, AcPase staining was rather light (Fig. 33).

In the caiman brain, the relative enzymatic activities in this nucleus were the reverse of those in turtles, for the neuronal somas were heavily stained for AcPase (Fig. 41), but most cells stained rather poorly for SDH (Fig. 40). The surrounding neuropil was somewhat more active for SDH than that of the turtle nucleus. In

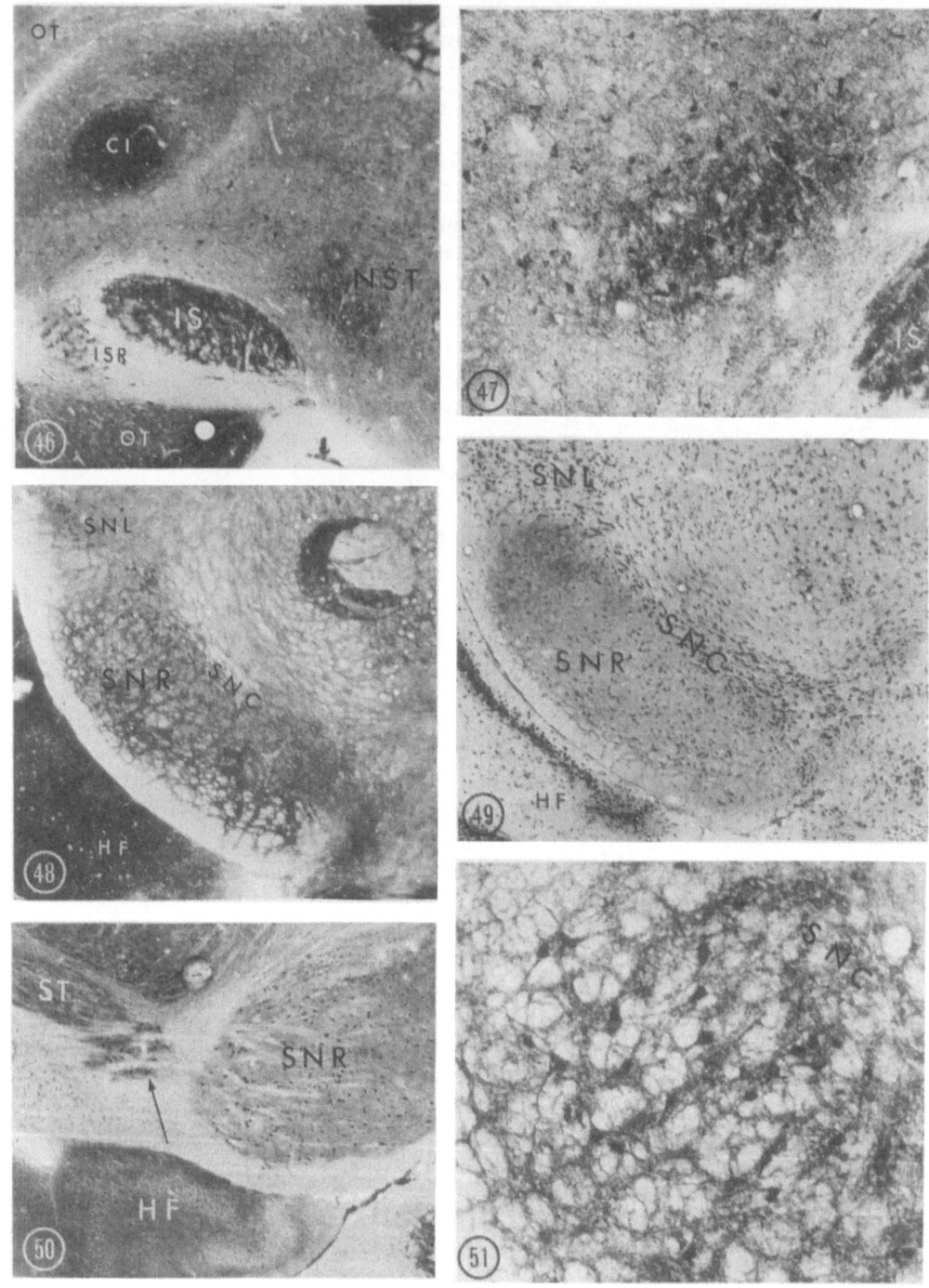

Fig. 46—51

Fig. 46. Parakeet. Coronal section, SDH. As seen here, the nucleus identified as CRAIGIE'S (1930) nucleus of the strio-tegmental tract, and the nucleus profundus mesencephali of the reptiles, show similar spatial relations to nearby structures. Like the nucleus profundus mesencephali of turtles

(continued page 61)

the two lizards, the cells of this nucleus shared the strong staining for SDH (and NAD diaphorase in *Anolis*) found in the turtle. Neuropil was moderately active in *Phrynosoma* and strongly so in *Anolis* (Figs. 43—45). In *Anolis*, the neurons of the nucleus profundus mesencephali were weakly stained for AcPase, but in *Phrynosoma*, they were highly active as in the caiman (Fig. 42).

Bird. Perhaps the most cogent evidence thus far presented for the homology of a structure in a submammalian brain with the substantia nigra of mammals is that of BERTLER *et.al.* (1964) and of FUXE and LJUNGGREN (1965). These investigators, using fluorescence histochemical techniques, found a large group of cells in the lateral midbrain tegmentum of the pigeon to have a high content of dopamine. These neurons were similar in this respect to those of the zona compacta of the substantia nigra of mammals. As in the case of the latter, these tegmental neurons in the pigeon brain were believed to form the origin of a dopamine pathway to the striatum. This cell group in the pigeon mesencephalon was believed by FUXE and LJUNGGREN to correspond at least in part with the nucleus tractus striotegmentalis described by CRAIGIE (1930) in the kiwi brain. CRAIGIE'S nucleus in this bird, and the substantia nigra identified in reptiles by HUBER and CROSBY (1933) are similar in their association with the strio-tegmental tract (see *Anolis* brain section illustrated in Fig. 10A of HUBER and CROSBY). CRAIGIE'S nucleus may correspond to the nucleus tegmenti pedunculo-pontinus pars compacta of KARTEN and HODOS (1967) in pigeon.

and lizards, the neurons of this avian nucleus are highly active for SDH. The surrounding neuropil also is quite strongly active. The structure in the parakeet brain which in these histochemical preparations clearly corresponds to the inferior colliculus of the reptiles, was called the nucleus mesencephalicus lateralis pars dorsalis by HUBER and CROSBY (1929). However, its homology with the inferior colliculus of mammals has been verified by KARTEN (1967) in the pigeon. × 35

Fig. 47. Parakeet. Coronal section, SDH. Section close to that shown in Fig. 46. At a higher magnification, the distribution of SDH activity in neurons and neuropil of the nucleus of the strio-tegmental tract is more clearly defined. × 100

Fig. 48. Mouse. Coronal section, SDH. Neuropil of the substantia nigra zona reticulata is moderately active for SDH. Some of the large, strongly stained neurons of this part of the substantia nigra are visible here, but are better visualized in Fig. 50 and 51. The neuropil of the zona compacta and pars lateralis is more lightly stained, and especially in the zona compacta the neuron bodies cannot be seen under low magnification, owing to their weak activity. Compare with Fig. 49. × 42.5

Fig. 49. Mouse. Coronal section, AcPase. In contrast to the activities for SDH seen in Figs. 48, 50 and 51, the large neurons of the zona reticulata are relatively lightly stained, whereas neurons of the zona compacta and pars lateralis are strongly so. × 42.5

Fig. 50. Mouse. Sagittal section, SDH. The substantia nigra zona reticulata shows relatively light staining in its neuropil and strong activity in its neurons. Rostral to it, among fiber bundles of the cerebral peduncle below the subthalamic nucleus, appear groups of strongly stained cells with rather active associated neuropil (arrows). These seem to represent a part of the entopeduncular nucleus which had moved caudally away from the remainder of that nucleus in this particular specimen. A broader view of this section appears in Fig. 32. × 45

Fig. 51. Mouse. Coronal section, SDH. Neurons of the substantia nigra zona reticulata show strong SDH activity in soma and processes. The surrounding neuropil is moderately stained. In the zona compacta, part of which appears at the upper right, the neurons are poorly visible even though the neuropil is lightly stained, owing to their low activity for this enzyme. × 180

In the present histochemical preparations of the parakeet brain, the nucleus identified as CRAIGIE's nucleus tractus striotegmentalis was prominent for dehydrogenase activity. The large neuron bodies were strongly stained, and the neuropil also was quite highly active, so that the cells and processes were partially obscured (Figs. 46, 47). In coronal sections, this nucleus was comparable to the nucleus profundus mesencephali of the reptile brains in its topographical relations to adjacent structures such as the optic tectum, inferior colliculus, and nucleus isthmi (Fig. 46). In the dehydrogenase activities studied, the avian nucleus was most like the nucleus profundus mesencephali of the lizards and turtle, and least like that of the caiman, with its poor activity. From incomplete preparations, the parakeet nucleus also appeared to differ from the nucleus profundus mesencephali of the caiman in showing a relatively lesser AcPase activity.

Mammals. In mammals, the term nucleus profundus mesencephali (and its equivalent, n. intrapeduncularis) was used by PAPEZ (1929, 1935) for a cell group dorsomedial to the substantia nigra, which he believed to form the true terminus of the strio-peduncular tract. This was noted by PAPEZ and ARONSON (1934) to be the probable homologue of the nucleus profundus mesencephali of reptiles. According to PAPEZ (1935), the very large nucleus profundus mesencephali of the anteater, *Tamandua* (ARIENS KAPPERS, 1921; KAELBER, 1966), corresponds to the nucleus so-designated in his own use of the term, but the name also has been used by other investigators for the lateral reticular mass of the midbrain as it is continued upwards from the pons.

A nucleus profundus mesencephali corresponding to PAPEZ' use of the term was not delineated in the rat brain by GILLILAN (1943) or by KONIG and KLIPPEL (1963). Nor, in the present histochemical study could such a cell group be identified in the mouse brain by comparison with PAPEZ and ARONSON's figures of its position in monkey brain. The large neurons and processes of the substantia nigra zona reticulata showed strong SDH activity (Figs. 50, 51), but were rather lightly stained for AcPase (Fig. 49). The neuropil of the zona reticulata was moderately or lightly stained for SDH, appearing paler in the more grossly fibrous rostral region (Fig. 50). In contrast to the neurons of the zona reticulata, those of the zona compacta were strongly stained for AcPase (Fig. 49), but poorly distinguishable from the lightly stained background neuropil in SDH preparations (Figs. 48, 51). Neuropil of both the zona compacta and pars lateralis was more lightly stained for SDH than that of the zona reticulata. Like cells of the zona compacta, neurons of the pars lateralis showed strong AcPase activity, and at least part of them were strongly stained for SDH. In general, the observations made on SDH activities in the various parts of the substantia nigra in the mouse agreed with those of FRIEDE (1959) in the guinea pig.

3. Interpeduncular Nucleus

This nucleus, in reptiles and mammals is basically similar in location, structure and connections. It consists mainly of a mass of neuropil, with closely aggregated very small cells largely dorsally located, and some larger cells laterally (GILLILAN, 1943). Histochemically, it was closely similar in all of the reptilian forms and in the mouse. For SDH, the neuropil probably was the most intensely stained area in any part of the brain, in all species (Figs. 35, 40, 43, 44). It also was heavily stained for NAD diaphorase (Fig. 37). No cell bodies could be distinguished in the dehydrogenase preparations, owing to the deep staining of the matrix. The small dorsal cells often were strongly stained for AcPase, but because of their tiny size, the overall effect in such preparations viewed at low magnification was unimpressive (Figs. 36, 41, 42).

In the parakeet, the interpeduncular nucleus showed relatively similar enzymatic activities to those shown by the nucleus in the reptiles and mammal, although the dehydrogenase activities were less striking. For these enzymes, this structure could not be said to be the most intensely stained part of the brain of the parakeet.

The two lizard species were unique in the possession of an habenulo-peduncular tract (fasciculus retroflexus) which was so intensely stained for SDH (and NAD diaphorase in *Anolis*) in routine incubations, as to appear almost black. Sagittal sections of the *Anolis* brain demonstrated this phenomenon most strikingly; in some sections a long extent of the tract could be seen passing over and down behind the nucleus rotundus of the thalamus (Fig. 39).

Discussion
Nucleus of the Dorsal Supraoptic Decussation

In comparing histochemical attributes in the region of the nucleus identifiable as the nucleus of the dorsal supraoptic decussation in the various reptiles studied, a pair of closely associated neuron groups was found which showed considerable parellelism in enzymatic activities and topographical relations from one species to another. In the turtle, it seemed likely that the two nuclei, although differentiable in oxidative enzyme activities, had been included as a single cytological entity in earlier descriptions of the nucleus of the dorsal supraoptic decussation (PAPEZ, 1935). The more rostral of these cell groups in the turtle was reticular in appearance, heavily stained in dehydrogenase preparations and was intercalated in the thalamic radiations. The more caudal group possessed a more homogeneous neuropil of moderate dehydrogenase activity, and seemed to be more closely associated with the ventral peduncle. These two nuclei are here referred to as "nucleus of the dorsal supraoptic decussation, rostral" and "nucleus of the dorsal supraoptic decussation, caudal".

In the dehydrogenase preparations of the lizard brains, two quite similar cell groups were seen, of which the more caudal one in *Phrynosoma* appeared to be linked by fibers with the supraoptic decussations. In the caiman brain dehydrogenase preparations, a large moderately stained caudal nucleus was definitely identified as HUBER and CROSBY's (1926) nucleus of the dorsal supraoptic decussation. An adjacent heavily stained rostral group in the caiman might have represented part of the interstitial nucleus of the olfactory projection tract. However, since this cell group as a histochemically differentiated entity could have represented but a small portion of the interstitial nucleus of the olfactory projection tract as delineated by HUBER and CROSBY, it seemed perhaps more likely that, as suggested above for the turtle, it represented a cell group included as part of the nucleus of the dorsal supraoptic decussation by HUBER and CROSBY. It might then represent a homologue of the rostral deeply stained reticular-structured nucleus associated with the dorsal peduncle in the turtle and lizards. At least for the sake of convenience, the same two names applied to the turtle nuclei are carried over to the caiman.

On the basis of position in relation to the dorsal peduncle of the lateral forebrain bundle, and on the basis of various histochemical resemblances, it is suggested that the more rostral nucleus in the reptiles might represent an homologue of the reticular nucleus of the thalamus in mammals. Support for this idea may be derived from HUBER and CROSBY's observation that "the nucleus of the dorsal supraoptic commissure receives collaterals and possibly stem fibers from the dorsal portion of the lateral forebrain bundle." Its definite interposition in the dorsal peduncle in turtle and lizards, and its at least close relation to the dorsal peduncular components in the caiman, would put it in the appropriate position to develop the kind of intrathalamic feedback mechanism and filtering of thalamic telencephalic output suggested by SCHEIBEL and SCHEIBEL (1967) for the nucleus reticularis thalami of mammals. The latter nucleus, in the cat, also receives fibers of the dorsal supraoptic decussation originating in the lower brain stem (BUCHER and BURGI, 1953).

The more caudal nucleus in the reptiles might then represent a nucleus of the dorsal supraoptic decussation in a more limited sense, and be homologous with that part of the entopeduncular nucleus of mammals which is related to the dorsal supraoptic decussations. Its seemingly closer relation to the ventral peduncle might favor such homology, and its histochemical differentiation could accord with the idea.

In the turtle, PAPEZ (1935) described an extensive nucleus suprapeduncularis, or reticularis, and concluded that it might correspond to the zone incerta and reticular nucleus of mammals, as "it is a bed nucleus of the thalamic radiation and of the bulbo-thalamic tract". He equated this with the area ventromedialis of HUBER and CROSBY (1926) in *Alligator*, and with part of the nucleus suprapeduncularis of FREDERIKSE (1931) in Lacerta. Both of the latter also were described as extensive areas, occupying most of the length of the ventral thalamus. The only structures showing special histochemical differentiation in the present study which might fall within these large areas were the more caudal nucleus associated above with the supraoptic decussations, and the nucleus referred to as "nucleus X", which was found only in the caiman and *Phrynosoma* brains. Both of these histo-chemically differentiated nuclei were much more circumscribed in extent than the areas described by the above-mentioned authors. "Nucleus X" in the caiman, however, might correspond to the rostral tip of HUBER and CROSBY's area ventro-medialis, whose first appearance rostrally they described as a "small round mass of cells", perhaps represented unlabelled in their Fig. 5.

Entopeduncular Nucleus

The entopeduncular nucleus of the reptile specimens showed some histo-chemical similarity to its suggested mammalian homologue, the subthalamic nucleus. This included moderate to strong AcPase and dehydrogenase staining of the neuron bodies and processes. The greatest difference was in the density of neuropil dehydrogenase activity. Since the subthalamic nucleus of the mammal is more segregated from the main stream of the ventral peduncular fibers than it is in the reptiles, it might be expected for this reason alone that its neuropil should appear more densely stained. The entopeduncular nucleus in the turtle and lizards appeared sparser in cell population and much less sharply defined than the subthalamic nucleus in the mouse, giving a general appearance of a considerably less differentiated state of development. However, in the caiman, the entopeduncular nucleus more closely resembled the subthalamic nucleus of the mouse in both enzyme staining and general appearance.

Red Nucleus

To a greater degree than the entopeduncular-subthalamic nucleus, the red nucleus in the reptiles showed poor histochemical activity and definition compared with that of mammals. From the point of view of the enzymatic activities studied, the emphasis in the ARIENS KAPPERS *et. al.* (1960) description of the red nucleus in reptiles should be on *"relatively"* (well-differentiated) and an italic added to HUBER and CROSBY's (1926) remark that evolutionary differentiation on the motor side of the reptile brain is "marked particularly by the *beginning* of a nucleus ruber . . .".

Nucleus Profundus Mesencephali: Substantia Nigra

There has been a great deal of variation in usage, in both reptiles and mammals, of the term "nucleus profundus mesencephali" (also see above, under Observations). In addition, there have been apparent inconsistencies in the identifications made of a substantia nigra homologue in reptiles. The latter was described by HUBER and CROSBY (1933) as a *large-celled* group receiving terminations of the ventral peduncle (see section in Observations). However, PAPEZ (1935) believed that a *small-celled* group dorsomedial to the large celled nucleus profundus mesencephali represented the substantia nigra in the turtle. He noted that the nucleus profundus mesencephali appeared to receive (presumably from the ventral peduncle) "a special bundle of fibers independent of those ending in the substantia nigra", and that the "nucleus of the substantia nigra" was closely related to the nucleus opticus tegmenti. He believed that dendrites of the neurons of the substantia nigra synapsed in the neuropil of the latter.

The nucleus which SHANKLIN (1930; 1933) called the "nucleus profundus mesencephali", and for which he described connections, clearly represents the rostral portion of the nucleus isthmi (or nucleus lateralis mesencephali — see HUBER and CROSBY, 1926). The nucleus isthmi of reptiles has been described as a part of the lateral lemniscal system by several authors (e.g. HUBER and CROSBY, 1926; LE GROS CLARK, 1932) (although doubt has recently been cast on this interpretation by KARTEN, 1967). Such overlapping of terminology may account for the statement of ARIENS KAPPERS *et al.* (1960) that the nucleus profundus mesencephali of reptiles is "essentially a nucleus of this" (lateral) "lemniscus". PAPEZ also noted that ARIENS KAPPERS "probably includes under this name also a more caudally situated cell group associated (in fishes) with the lateral lemniscus."

Terms used in describing a nucleus apparently homologous with the reptilian nucleus profundus mesencephali, or with the mammalian substantia nigra, by students of the avian brain were noted above under Observations.

The nucleus profundus mesencephali in the several reptile species included in the present study showed differential histochemical resemblances to parts of the substantia nigra of mammals. That in the caiman was similar to the zona compacta and pars lateralis of the mouse in showing intense AcPase activity in the neurons, and also resembled the zona compacta of mouse and other mammals in the poor dehydrogenase activity of its cells. That in the turtle and *Anolis* resembled more the zona reticulata of the mouse and other mammals in its strong neuronal dehydrogenase staining, and also resembled the zona reticulata of the mouse in the weak to moderate AcPase activity in its cells. In *Phrynosoma*, the nucleus profundus mesencephali combined strong staining for both AcPase and SDH in its neurons, thus sharing histochemical characters with all parts of the mammalian substantia nigra. Neuropil surrounding the cells of the nucleus profundus mesencephali was weakly or moderately active for the dehydrogenases in caiman, turtle and *Phrynosoma*, but was strongly so in *Anolis*. The latter feature appears not to be characteristic of any part of the mammalian substantia nigra as reported in the histochemical literature.

In the parakeet brain, the nucleus tractus strio-tegmentalis, here concluded to represent an homologue of the nucleus profundus mesencephali of the reptiles, was most like that of *Anolis* in the activity of its neurons and neuropil for the enzymes tested. That it differed most from the nucleus profundus mesencephali of the caiman is perhaps surprising in view of the close taxonomic relationship of birds and Archosaurians. As noted above under Observations, the nucleus of the strio-tegmental tract included neurons whose biogenetic amine histochemistry paralleled neurons of the substantia nigra zona compacta of mammals. Hence, if the comparative identifications made here are correct, this nucleus in the avian brain was similar to the mammalian zona compacta in dopamine metabolism, but more like the mammalian zona reticulata with respect to the neurologically non-specific enzyme activities described here.

In view of the difficulty in pinning down the identifications and comparisons of nuclei in this lateral midbrain area which have been made by various students of the reptilian and avian brains, it would be of importance to study monoamine metabolism in this region in a comparative series of reptiles and birds. Greater cytological resolution of the enzyme activities reported on here would also be desirable, as it could not be ascertained from the present material whether the same population of neurons in the nucleus profundus mesencephali of *Phrynosoma* or turtle was being demonstrated in the dehydrogenase versus acid phosphatase incubates. Further biochemical and physiological study is plainly necessary before any definite parallels can be drawn between structures in submammalian brains and the substantia nigra of mammals.

Summary

1. Comparative histochemical studies were made, by the application of methods for the demonstration of the activities of four enzymes, on basal brainstem structures of several species of reptiles, a bird and a mammal. The animal specimens included turtles, young Spectacled Caimans, Texas Horned Lizards, Carolina Anoles, Shell Parakeets and domestic mice. The enzyme activities chiefly studied

were succinate dehydrogenase, nicotinamide-adenine dinucleotide tetrazolium reductase and acid phosphatase. A method for demonstration of thiamine pyrophosphatase activity was applied to brains of some of the species.

2. In the reptile specimens, two contiguous nuclei with differentiating enzymatic characteristics were found in close association with components of the lateral forebrain bundle in the rostral subthalamus. One or both of these had been described in the literature as the nucleus of the dorsal supraoptic decussation. On the basis of comparative histochemical characteristics, as well as topographical relations, it was suggested that one of these nuclei might be comparable to the reticular nucleus of the thalamus in mammals, and the other to part of the entopeduncular nucleus of mammals.

3. The reptilian entopeduncular nucleus showed considerable histochemical resemblance to its suggested mammalian homologue, the subthalamic nucleus. This resemblance was most marked in the caiman.

4. The red nucleus of the reptiles was poorly differentiated with respect to the enzymatic activities tested in comparison with that of mammals.

5. The reptilian nucleus profundus mesencephali was equated with the nucleus tractus strio-tegmentalis of the parakeet. Both were compared with the mammalian substantia nigra, as the reptilian nucleus appeared to correspond to certain nuclei described in the literature as representing the substantia nigra of reptiles, and as the nucleus in the bird appeared to correspond to that reported to show a dopamine metabolism similar to that of the zona compacta of the substantia nigra of mammals. In the mouse, the zona reticulata and zona compacta of the substantia nigra had opposing enzymatic characteristics: strong acid phosphatase and low dehydrogenase activities in the zona compacta, and the reverse in the zona reticulata. The nucleus profundus mesencephali in the turtles was similar in its enzymatic activities to the zona reticulata of mammals, but in the caiman the enzymatic activities of the nucleus were like those of the zona compacta of mammals. The nucleus profundus mesencephali of the anoles and the nucleus tractus strio-tegmentalis of the parakeet showed most resemblance to the zona reticulata, although differing to some extent in the strong dehydrogenase activity seen in the associated neuropil. The nucleus profundus mesencephali of the horned lizards showed a combination of both types of enzyme activity patterns.

6. Enzymatic characteristics of the interpeduncular nucleus were similar in all species.

Acknowledgments. The encouragement and assistance of the same persons cited in the first part of this study is gratefully acknowledged. In addition, thanks are due to M. DANIEL PONTHIEUX for photographic help in the preparation of the present paper.

Bibliography

ARIENS KAPPERS, C. U.: Die vergleichende Anatomie des Nervensystems der Wirbeltiere und des Menschen. Haarlem: E. F. Bohn 1921.
— C. G. HUBER, and E. C. CROSBY: The comparative anatomy of the nervous system of vertebrates, including man. New York: Hafner, Reprint Ed., 1960.
BAKER-COHEN, K. F.: Comparative enzyme histochemical observations on submammalian brains. I. Striatal structures in reptiles and bird. Ergebn. Anat. Entwickl.-Gesch. 1968.
BECCARI, N.: Il centro tegmentale o interstiziale ed altre formazione poco nete nel mesencefalo e nel diencefalo di un rettile. Arch. ital. Anat. Embriol. **20**, 560—619 (1923).

BERTLER, Å., B. FALCK, C. G. GOTTFIRES, L. LJUNGGREN, and E. ROSENGREN: Some observations on adrenergic connections between mesencephalon and cerebral hemispheres. Acta pharmacol. (Kbh.) **21**, 283—289 (1964).

BUCHER, V. M., and S. E. BURGI: Some observations on the fiber connections of the di- and mesencephalon in the cat. III. The supraoptic decussations. J. comp. Neurol. **98**, 355—379 (1953).

CRAIGIE, E. H.: Studies on the brain of the kiwi. J. comp. Neurol. **49**, 223—357 (1930).

DAHLSTRÖM, A., and K. FUXE: Evidence for the existence of monoamine-containing neurons in the central nervous system. I. Demonstration of monoamines in the cell bodies of brain stem neurons. Acta physiol. scand. **62**, Suppl. 232, 1—55 (1965a).

— — Evidence for the existence of monoamine neurons in the central nervous system. IV. Distribution of monoamine nerve terminals in the central nervous system. Acta physiol. scand. **64**, Suppl. 247, 37—85 (1965b).

DE LANGE, S. J.: Das Zwischenhirn und das Mittelhirn der Reptilien. Folia neuro-biol. (Lpz.) **7**, 67—138 (1913).

DURWARD, A.: The cell masses in the forebrain of *Sphenodon punctatum*. J. Anat. (Lond.) **65**, 8—44 (1930).

FALCK, B.: Observations on the possibilieies of the cellular localization of monoamines by a fluorescence method. Acta physiol. scand. **56**, Suppl. 197, 1—25 (1962).

— Cellular localization of monoamines. In: Progress in brain research, biogenetic amines (H. E. HIMWICH and W. A. HIMWICH, eds.), vol. 8, p. 28—44. Amsterdam: Elsevier 1964.

FREDERICKSE, A.: The lizard's brain. An investigation on the histological structure of the brain of *Lacerta vivipara*. Baarn/The Netherlands: C. C. Callenbach 1931.

FRIEDE, R. L.: Histochemical investigations on succinic dehydrogenase in the central nervous system. III. Atlas of the midbrain of the guinea pig, including pons and cerebellum. J. Neurochem. **4**, 290—303 (1959).

— Histochemical investigations on succinic dehydrogenase in the central nervous system. V. The diencephalon and basal telencephalic centres of the guinea pig. J. Neurochem. **6**, 190—199 (1961a).

— A histochemical atlas of tissue oxidation in the brain stem of the cat. New York: Hafner (1961b.

— Topographic brain chemistry. New York: Academic Press 1966.

FUXE, K., and L. LJUNGGREN: Cellular localization of monoamines in the upper brainstem of the pigeon. J. comp. Neurol. **125**, 355—381 (1965).

GILLILAN, L. A.: The nuclear pattern of the non-tectal portions of the midbrain and isthmus in rodents. J. comp. Neurol. **78**, 213—251 (1943).

HUBER, G. C., and E. C. CROSBY: On thalamic and tectal nuclei and fiber paths in the brain of the American alligator. J. comp. Neurol. **40**, 97—227 (1926).

— The reptilian optic tectum. J. comp. Neurol. **57**, 57—163 (1933).

JOHNSTON, J. B.: Further contributions to the study of the evolution of the forebrain. J. comp. Neurol. **35**, 337—481 (1923).

KAELBER, W. W.: Nuclear configuration of the diencephalon of *Tamandua tetradactyla* and *Myrmecophaga jubata*. J. comp. Neurol. **128**, 133—170 (1966).

KARTEN, H. J.: The organization of the ascending auditory pathway in the pigeon (*Columba livia*). I. Diencephalic projections of the inferior colliculus (nucleus mesencephali lateralis, pars dorsalis). Brain Res. **6**, 409—427 (1967).

and W. HODOS: A stereotaxic atlas of the brain of the pigeon (*Columba livia*). Baltimore: The Johns Hopkins Press 1967.

KONIG, J. F. R., and R. A. KLIPPEL: The rat brain. A stereotaxic atlas. Baltimore: Williams & Wilkins Co. 1963.

LE GROS CLARK, W. E.: The medial geniculate body and nucleus isthmi. J. Anat. (Lond.) **67**, 536—548 (1932).

NAUTA, W. J. H., and W. R. MEHLER: Some efferent connections of the lentiform nucleus in monkey and cat. Anat. Rec. **139**, 260 (1961).

PAPEZ, J. W.: Thalamus of turtles and thalamic evolution. J. comp. Neurol. **61**, 433—475 (1935).

— Comparative neurology, reprint of 1929 publication. New York: Hafner 1961.

PAPEZ, J. W., and L. ARONSON: Thalamic nuclei of *Pithecus (Macacus) rhesus*. I. Ventral thalamus. Arch. Neurol. Psychiat. (Chic.) **32**, 1—26 (1934).

RIOCH, D. McK.: Studies on the diencephalon of carnivora. II. Certain nuclear configurations and fiber connections of the subthalamus and midbrain of the dog and cat. J comp. Neurol. **49**, 121—153 (1929).

SCHARRER, E.: Functional organization of the brain. In: Neuropharmacology (H. A. ABRAMSON, ed.), p. 90—106. New York: Josiah Macy, Jr. Foundation 1955.

SCHEIBEL, M. E., and A. B. SCHEIBEL: Structural organization of nonspecific thalamic nuclei and their projection toward cortex. Brain Res. **6**, 60—94 (1967).

SHANKLIN, W. M.: The central nervous system of *Chamaeleon vulgaris*. Acta zool. (Stockh.) **11**, 425—490 (1930).

— The comparative neurology of the nucleus opticus tegmenti with special reference to *Chamaeleon vulgaris*. Acta zool. (Stockh.) **14**, 163—184 (1933).

Subject Index

Alligator   13, 35, 47, 49, 57, 59, 64
Amygdala   16, 17, 22—29
Amygdaloid ridge   11
Anterior olfactory nucleus   11
Archistriatum   10, 23, 25, 37

Caudate nucleus   11—15, 18, 32—35, 37, 38
Cerebellum   28, 29, 52
Chameleon   10, 35, 59
Chelone midas   57
Chemoarchitectonics   7, 39, 42
Chicken brain   14
Choroid plexus   20
Claustrum   33
Core nucleus   15—18, 20—22, 34, 35, 38

Dorsolateral area   13, 16—19, 24, 25, 36—38, 50, 52
Draco volans   57

Ectostriatum   14, 16, 17, 25, 28—31, 34, 35, 38
Entopeduncular nucleus (of mammals)   42, 52, 53, 64, 67
— — (of reptiles)   50—53, 57, 58, 64, 65, 67

Fasciculus retroflexus   54, 55, 63
Fluorescence histochemistry   8, 35, 42, 61, 66
Fornix   50, 51
Frog brain   8

Globus pallidus   11—14, 16, 17, 20, 21, 24—38, 43—45, 52, 53
Glycogen   8
Golgi apparatus   32

Habenula   24, 25, 48—51
Hippocampal formation   16, 17, 21—31, 46, 47, 52, 53, 60, 61
Hyperstriatum   8, 16, 25, 27—29, 33, 34, 37, 39
Hyperstriatum accessorium   14, 16, 28, 29
Hypothalamus   24, 25, 46—48, 50—52

Inferior colliculus   28, 29, 52—57, 60, 61
Intermediolateral area   13, 14, 16, 19, 23—27, 36—38
Interpeduncular nucleus   22—25, 54—58, 62, 63, 67

Kiwi brain   61

Lacerta   10, 59, 64
Lateral forebrain bundle   11—13, 16, 17, 20—29, 33, 36, 42—53, 58, 59
Lateral geniculate nucleus   46, 47
Lateral olfactory tract, and nucleus of, 11, 16—18, 24, 25, 30, 34
Lentiform nucleus   10, 11, 13
Lysosomes   32

Mesencephalic V nucleus   8
Mitochondria   31

Neostriatum   11, 28—35, 37, 39
Nucleus accumbens   33
Nucleus anterior dorsalis   30, 31
Nucleus dorsalis commissure posterior   52
Nucleus dorsolateralis anterior   48, 49
Nucleus isthmi   17, 24, 25, 28, 29, 54—57, 60, 61, 65
Nucleus of the basal optic root   24, 25
— of the dorsal supraoptic decussation   43—57, 63, 64, 67

Nucleus ovalis   46, 47
Nucleus profundus mesencephali (mammals)
    62
— — — (reptiles)   54—59, 61, 62, 65
Nucleus reticularis thalami   52, 53, 64, 67
Nucleus reuniens   24, 25, 44, 45, 50, 51
Nucleus rotundus   22—25, 28, 29, 36,
    43—53
Nucleus tegmenti pedunculo-pontinus pars
    compacta   61
Nucleus tractus striotegmentalis   60—62,
    66

Oculomotor nucleus   22—25, 54—59
Optic tectum   24, 25, 28, 29, 36, 50—57, 60

Palaeostriatum   10—12, 32—34, 37, 38
Palaeostriatum augmentatum   14, 16, 17,
    24—31, 34, 35, 37, 38
Palaestriatum primitivum   14
Pallial thickening   21, 34, 35
Pigeon brain   8, 35, 36, 61
Piriform lobe, and cortex   10, 16—18,
    20—22, 34
Pretectal nucleus   28, 29, 50, 51
Primate brain   55

Primordial general cortex   35
Putamen   11—18, 20, 21, 28—31, 33—38,
    44, 45, 52

Red nucleus   54, 55, 58, 59, 65, 67

Selachian brain   8
Sphenodon   10, 12, 43
Stria bed   11, 33
Substantia nigra   42, 58, 65, 66
— —, pars lateralis   60—62
— —, zona compacta   60—62, 67
— —, zona reticulata   52, 53, 60—62, 67
Subthalamic nucleus   42, 52, 53, 58, 60,
    61, 64, 65, 67

Tamandua   62
Tuberculum olfactorium   30, 31

Ventral thalamic nucleus   36
Ventrolateral large-celled area   13, 14, 25
Ventrolateral small-celled area   13, 14, 16,
    34, 38

Zinc   8
Zona incerta   52, 53, 64

Ergebnisse der Anatomie
und Entwicklungsgeschichte

Advances in Anatomy
Embryology and Cell Biology

Revues d'anatomie
et de morphologie expérimentale

Editores

A. Brodal, Oslo · W. Hild, Galveston · R. Ortmann, Köln
T. H. Schiebler, Würzburg · G. Töndury, Zürich · E. Wolff, Paris

Band 40 (Heft 1—6)

Springer-Verlag Berlin · Heidelberg · New York 1968

Inhalt

Heft 1: **Die extracutanen Melanocyten der Echsen (Sauria)**

Von HANS-RAINER DUNCKER

Heft 2: **Glykogen in der Ontogenese des Verdauungstrakts**
Chemomorphologische und stoffwechselhistochemische Analyse

Von DIETER SASSE

Heft 3: **Die Muskelanordnung in der Speiseröhre**
(Mensch, Rhesusaffe, Kaninchen, Maus, Ratte, Seehund)

Von PETER KAUFMANN, WERNER LIERSE, JOCHEN STARK
und FRIEDRICH STELZNER

Heft 4: **Die Angioarchitektur im Oesophagus des Kaninchens, der Ratte und der Maus**

Von STEFAN GÜNTHER und WERNER LIERSE

Heft 5: **Intrinsic Neuronal Organization of the Vestibular Nuclear Complex in the Cat.** A Golgi Study

By EIVINN HAUGLIE-HANSSEN

Heft 6: **Comparative Enzyme Histochemical Observations on Submammalian Brains**

Part I. Striatal Structures in Reptiles and Birds
Part II. Basal Structures of the Brainstem in Reptiles and Bird

By K. FRANCE BAKER-COHEN